ExamKrackers MCAT

VERBAL REASONING AND MATH

5TH EDITION

Published in the United States of America by Osote Publishing, New Jersey

ISBN 1-893858-35-9 (Volume 2)
ISBN 1-893858-36-7 (5 Volume Set)

5th Edition

To purchase additional copies of this book or the rest of the 5 volume set,
call 1-888-572-2536 or fax orders to 1-201-797-1644.

examkrackers.com

osote.com

audioosmosis.com

Inside layout design: Saucy Enterprizes (www.saucyenterprizes.com)
Cover design: Scott Wolfe
Inside cover design consultant: Fenwick Design Inc. (212) 246-9722; (201) 944-4337
Illustrations by David Orsay and the ExamKrackers staff.

Printed and Bound in China

Acknowledgements

Although I am the author, the hard work and expertise of many individuals contributed to this book. The idea of writing in two voices, a science voice and an MCAT voice, was the creative brainchild of my imaginative friend Jordan Zaretsky. I would like to thank David Orsay for his help with the verbal passages. I wish to thank my wife, Silvia, for her support during the difficult times in the past and those that lie ahead.

Finally, I wish to thank my daughter Julianna Orsay for helping out whenever possible.

Table of Contents

intro i

Introduction to MCAT Including MCAT Math

i-1 The Layout of the MCAT

The MCAT consists of four sections:

1. The Physical Sciences Section
2. The Verbal Reasoning Section

   Lunch

3. The Writing Sample Section
4. The Biological Sciences Section

The Physical Sciences Section is 100 minutes and covers science topics from first year undergraduate physics and inorganic chemistry. It consists of 77 multiple-choice questions with answer choices A through D. There are 11 short passages followed by 4 to ten questions each. Passages average approximately 200 words in length and are often accompanied by one or more charts, diagrams, or tables. There are 3 sets of stand-alone multiple-choice questions dispersed throughout the test as well. The top score on the Physical Sciences Section is a 15.

There is a 10 minute break following the Physical Sciences Section.

The Verbal Reasoning Section is 85 minutes. It consists of 60 multiple-choice questions with answer choices A through D. There are 9 passages followed by 4 to ten questions each. Passages average approximately 600 words in length. There is a wide variety of passage topics ranging from economics and anthropology to poetic analysis. The top score on the Verbal Reasoning Section is a 15.

There is a 60 minute lunch break following the Verbal Reasoning Section.

The Writing Sample Section is two 30 minute periods. No break is given between the periods. In each period the test-taker is given an argument in the form of a statement and asked to take both sides of the argument and give a guideline to determine which side is right in which case. The top score on the Writing Sample Section is a T on a alphabetic scale from J to T. This scale translates to a score of 1-6 on each essay resulting in a combined score of a 2-12 represented by J through T.

The Biological Sciences Section is 100 minutes and covers science topics from a wide range of first year undergraduate biology topics, organic chemistry and genetics. It consists of 77 multiple-choice questions with answer choices A through D. There are 11 short passages followed by 4 to ten questions each. Passages average approximately 200 words in length and are often accompanied by one or more charts, diagrams, or tables. There are 3 sets of stand-alone multiple-choice questions dispersed throughout the test as well. The top score on the Biological Sciences Section is a 15.

i-2 The Writing Sample

**The section that follows includes material from the MCAT Practice Test III. These materials are reprinted with permission of the Association of American Medical Colleges (AAMC).

In the U.S., your writing sample score is unlikely to affect whether or not you gain admittance to medical school. Currently, medical schools do not give this section much weight in their decision making process. Medical schools do not see your actual writing sample. They only see your score. The writing sample functions to wear you down for the Biological Sciences Section.

The writing sample is more of an exercise in following directions than it is a test of your ability to write. You should not attempt to be creative on the writing sample or try to make your reader reflect deeply. Instead, follow the simple three step process given below. Two sample statements are given with each step followed by an example of how your essay should appear for that statement.

A similar set of directions is always given with each statement. Don't waste time reading the directions on the real MCAT. The directions can be summarized into the following **three step process**:

1. **Explain the statement as thoroughly as possible using an example to clarify.**

   Statement: *An understanding of the past is necessary for solving the problems of the present.* **Paraphrase:** *History is an integral part of the learning process. By studying the past, we can analyze repercussions of certain behavior and action patterns.*

   Statement: *No matter how oppressive a government, violent revolution is never justified.* **Paraphrase:** *The familiar idiom "He who lives by the sword shall die by the sword", is echoed in any statement that condemns violence. It is a very simple principle based on a very logical argument. Violence invites more of the same. If a government is overthrown by violent means, then there is a precedent set and there is nothing stopping others from doing the same again.*

   Do not: *begin your essay with the statement "so and so" means that…*

2. **Give a specific example contradicting the statement.**

   Statement: *An understanding of the past is necessary for solving the problems of the present.* **Example:** *On the other hand, some problems exist today that are totally independent of any historical event. The current issue of AIDS…*

   Statement: *No matter how oppressive a government, violent revolution is never justified.* **Example:** *However, there can be times when extreme action becomes necessary. It was the violence of the Russian revolution that brought an end to the suffering of the masses during WWI.*

   Do not: *use controversial topics as examples, such as abortion or contemporary political issues.*

3. **Give a guideline that anyone might use to determine when the statement is true and when it is false.**

   Statement: *An understanding of the past is necessary for solving the problems of the present.* **Guideline:** *When then is the past crucial to our understanding of the current events? It is important only, and especially, when it relates to the present situation….*

> **Statement:** *No matter how oppressive a government, violent revolution is never justified.* **Guideline:** *Whether or not violent revolution is justified depends upon whether some form of oppression is lifted from the masses.*

Spend the first 5 minutes of the essay writing an **outline** of these three steps.

Write **2 pages**. Be sure to **finish your essay**. The outline should help you do this. Above all, **write neatly**. Use proper grammar correctly. Don't misspell words. Don't use words if you are not certain of the meaning. Historical examples are much better than personal examples; "Martin Luther King said..." is a much better example than "My mother always said..." To think of examples, think of wars or famous people. Feel free to paraphrase liberally: "Socrates once said that he was the smartest man because he understood how little he really knew." This is not an accurate quote; it is an acceptable paraphrase. Socrates said something like this, and this is close enough.

---

i-3 How to Approach Science Passages

The following guidelines should be followed when working an MCAT science passage:

1. **Read the passage first.** Regardless of your level of science ability, you should read the passage. Passages often give special conditions that you would have no reason to suspect without reading and which can invalidate an otherwise correct answer.
2. **Read quickly; do not try to master the information given in the passage.** Passages are full of information both useful and irrelevant to the adjoining questions. Do not waste time by attempting to gain complete understanding of the passage.
3. **Quickly check tables, graphs, and charts.** Do not spend time studying tables, graphs, and charts. Often, no questions will be asked concerning their content. Instead, quickly check headings, titles, axes, and obvious trends.
4. **When multiple hypotheses or experiments are posited, make note of obvious contrasts in the margin alongside the respective paragraphs.** Making note in the margin will accomplish two things. First, it will distinguish firmly in your mind each of the hypotheses or experiments. (At least one question will require such discernment.) Second, by labeling them you prevent confusion and thus obviate rereading (and avoid wasting precious time).
5. **Pay close attention to detail in the questions.** The key to a question is often found in a single word, such as "net force" or "constant velocity".
6. **Read answer choices immediately, before doing calculations.** Answer choices give information. Often a question that appears to require extensive calculations can be solved by intuition or estimation due to limited reasonable answer choices. Sometimes answer choices can be eliminated for having the wrong units, being nonsensical, or other reasons.
7. **Fill in your answer grid question by question as you go.** This is the best way to avoid bubbling errors. This method avoids time wasted trying to find your place. The posited reason for doing differently is that you can relax your brain while you transfer your answers. Try it. It's not relaxing. In fact, if you do relax, you are likely to make errors.

8. **If time is a factor for you, skip the questions and/or passages that you find difficult.** If you usually do not finish this section, then make sure that you at least answer all of the easy questions. In other words, guess at the difficult questions and come back to them if you have time. Be sure to make time to answer all of the free-standing questions. The free-standing questions are usually easier than those based on passages. By the time you have finished this course, you should not need to skip any questions.

i-4 MCAT Math

MCAT math will not test your math skills beyond the contents of this book. The MCAT *does* require knowledge of the following up to a second year high school algebra level: ratios; proportions; square roots; exponents and logarithms; scientific notation; quadratic and simultaneous equations; graphs. In addition, the MCAT tests: vector addition, subtraction; basic trigonometry; very basic probabilities. The MCAT *does not* test dot product, cross product or calculus.

This is me after I hurt myself with complicated calculations on my first MCAT.

Calculators are neither allowed on the MCAT, nor would they be helpful. From this moment until MCAT day, you should do all math problems in your head whenever possible. Do not use a calculator, and use your pencil as seldom as possible, when you do any math.

If you find yourself doing a lot of calculations on the MCAT, it's a good indication that you are doing something wrong. As a rule of thumb, **spend no more than 3 minutes on any MCAT physics question**. Once you have spent 3 minutes on a question with no resolution, you should stop what you're doing and read the question again for a simple answer. If you don't see a simple answer, you should make your best guess and move to the next question.

i-5 Rounding

Exact numbers are rarely useful on the MCAT. In order to save time and avoid errors when making calculations on the test, **use round numbers**. For instance, the gravitational constant **g should be rounded up to 10 m/s²** for the purpose of calculations, even when instructed by the MCAT to do otherwise. Calculations like 23.4×9.8 should be thought of as "something less than 23.4×10, which equals something less than 234 or less than 2.34×10^2." Thus if you see a question requiring the above calculations followed by these answer choices:

A. 1.24×10^2
B. 1.81×10^2
C. 2.28×10^2
D. 2.35×10^2

2.34
× 9.8
18.72
210.60
229.32

Wrong way

Answer is something less than $23.4 \times 10 = 234$.

Right way

answer choice C is the closest answer under 2.34×10^2, and C should be chosen quickly without resorting to complicated calculations. Rarely will there be two possible answer choices close enough to prevent a correct selection after rounding. If two answer choices on the MCAT are so close that you find you have to write down the math, it's probably because you've made a mistake. If you find yourself in that situation, look again at the question for a simple solution. If you don't see it, guess and go on.

It is helpful to **remain aware of the direction in which you have rounded**. In the above example, since answer choice D is closer to 234 than answer choice C, you may have been tempted to choose it. However, a quick check on the direction of rounding would confirm that 9.8 was rounded upward so the answer should be less than 234. Again, assuming the above calculations were necessary to arrive at the answer, an answer choice which would prevent the use of rounding, like 2.32×10^2 for instance, simply would not appear as an answer choice on a real MCAT. It would not appear for the very reason that such an answer choice would force the test taker to spend time making complicated calculations, and those aren't the skills the MCAT is designed to test.

If a series of calculations is used where rounding is performed at each step, the rounding errors can be compounded and the resulting answer can be useless. For instance, we may be required to take the above example and further divide "23.4×9.8" by 4.4. We might round 4.4 down to 4, and divide 240 by 4 to get 60; however, each of our roundings would have increased our result compounding the error. Instead, it is better to round 4.4 up to 5, dividing 235 by 5 to get 47. This is closer to the exact answer of 52.1182. In an attempt to increase the accuracy of multiple estimations, **try to compensate for upward rounding with downward rounding in the same calculations**.

Notice, in the example, that when we increase the denominator, we are decreasing the entire term. For instance:

$$\frac{625}{24} = 26.042 \qquad \frac{625}{25} = 25$$

Rounding 24 up to 25 results in a decrease in the overall term.

When rounding squares remember that you are really rounding twice. $(2.2)^2$ is really 2.2×2.2, so when we say that the answer is something greater than 4 we need to keep in mind that it is significantly greater because we have rounded down twice. One way to increase your accuracy is to round just one of the 2.2s, leaving you with something greater than 4.4. This is much closer to the exact answer of 4.84.

Another strategy for rounding an exponential term is to remember that difficult-to-solve exponential terms must lie between two easy-to-solve exponential terms. Thus 2.2^2 is between 2^2 and 3^2, closer to 2^2. This strategy is especially helpful for square roots. The square root of 21 must be between the square root of 16 and the square root of 25. Thus, the MCAT square root of 21 must be between 5 and 4 or about 4.6.

$$\sqrt{25} = 5$$
$$\sqrt{21} = ?$$
$$\sqrt{16} = 4$$

For more complicated roots, recall that any root is simply a fractional exponent. For instance, the square root of 9 is the same as $9^{1/2}$. This means that the fourth root of 4 is $4^{1/4}$. This is the same as $(4^{1/2})^{1/2}$ or $\sqrt{2}$. We can combine these techniques to solve even more complicated roots:

$$\sqrt[3]{27} = 3$$

$$4^{\frac{2}{3}} = \sqrt[3]{4^2} = \sqrt[3]{16} = ? = 2.51$$

$$\sqrt[3]{8} = 2$$

It's worth your time to memorize $\sqrt{2} \approx 1.4$ and $\sqrt{3} \approx 1.7$.

The MCAT is likely to give you any values that you need for trigonometric functions; however, since MCAT typically uses common angles, it is a good idea to be familiar with trigonometric values for common angles. Use the paradigm below to remember the values of common angles. Notice that the sine values are the reverse of the cosine values. Also notice that the numbers under the radical are 0, 1, 2, 3 and 4 from top to bottom for the sine function and bottom to top for the cosine function, and all are divided by 2.

θ	sine	cosine
0°	$\frac{\sqrt{0}}{2}$	$\frac{\sqrt{4}}{2}$
30°	$\frac{\sqrt{1}}{2}$	$\frac{\sqrt{3}}{2}$
45°	$\frac{\sqrt{2}}{2}$	$\frac{\sqrt{2}}{2}$
60°	$\frac{\sqrt{3}}{2}$	$\frac{\sqrt{1}}{2}$
90°	$\frac{\sqrt{4}}{2}$	$\frac{\sqrt{0}}{2}$

Less practiced test takers may perceive a rounding strategy as risky. On the contrary, **the test makers actually design their answers with a rounding strategy in mind**. Complicated numbers can be intimidating to anyone not comfortable with a rounding strategy.

Questions:

Solve the following problems by rounding. Do not use a pencil or a calculator.

1. $\dfrac{5.4 \times 7.1 \times 3.2}{4.6^2}$

   A. 2.2
   B. 3.8
   C. 5.8
   D. 7.9

2. $\dfrac{\sqrt{360 \times 9.8}}{6.2}$

   A. 9.6
   B. 13.2
   C. 17.3
   D. 20.2

3. $\dfrac{(\sqrt{2}) \times 23}{50}$

   A. 0.12
   B. 0.49
   C. 0.65
   D. 1.1

4. $\dfrac{(2 \times 45)^2}{9.8 \times 21}$

   A. 11
   B. 39
   C. 86
   D. 450

5. $\sqrt{\dfrac{2 \times 9.8^2}{49}}$

   A. 0.3
   B. 0.8
   C. 1.2
   D. 2

Answers:

1. **C is correct.** The exact answer is 5.7981.
2. **A is correct.** The exact answer is 9.5802.
3. **C is correct.** The exact answer is 0.65054.
4. **B is correct.** The exact answer is 39.359.
5. **D is correct.** The exact answer is 1.9799.

i-6 Scientific Notation

One important math skill tested rigorously by the MCAT is your ability to use scientific notation. In order to maximize your MCAT score, you must be familiar with the techniques and shortcuts of scientific notation. Although it may not seem so, scientific notation was designed to make math easier, and it does. You should practice the following techniques until you come to view scientific notation as a valuable ally.

This manual will define the terms in scientific notation as follows:

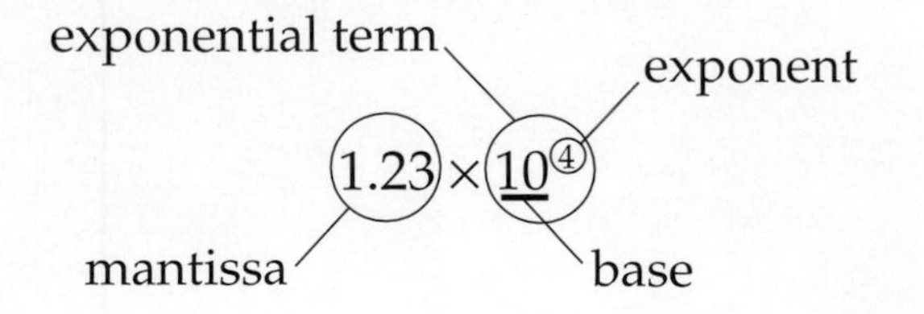

Magnitude: You should try to gain a feel for the exponential aspect of scientific notation. 10^{-8} is much greater than 10^{-12}. It is 10,000 times greater! Thus, when comparing one solution whose concentration of particles is 3.2×10^{-11} mol/L with a second solution whose concentration of particles is 4.1×10^{-9} mol/L, you should visualize the second solution as hundreds of times more concentrated than the first. Pay special attention to magnitudes when adding. For example try solving:

$$\begin{array}{r} 3.74 \times 10^{-15} \\ + \; 6.43 \times 10^{-3} \\ \hline \end{array}$$

On the MCAT, the answer is simply 6.43×10^{-3}. This is true because 6.43×10^{-3} is so much greater than 3.74×10^{-15} that 3.74×10^{-15} is negligible. Thus you can round off the answer to 6.43×10^{-3}. After all, the exact answer is 0.00643000000000374. Try solving:

$$\begin{array}{r} 5.32 \times 10^{-4} \\ \times \; 1.12 \times 10^{-13} \\ \hline \end{array}$$

The MCAT answer is something greater than 5.3×10^{-17}. We cannot ignore the smaller number in this case because we are multiplying. **In addition or subtraction, a number 100 times smaller or more can be considered negligible. This is not true in multiplication or division**.

The fastest way to add or subtract numbers in scientific notation is to make the exponents match. For instance:

$$\begin{array}{r} 2.76 \times 10^{4} \\ + \; 6.91 \times 10^{5} \\ \hline \end{array}$$

The MCAT answer is something less than 7.2×10^{5}. To get this answer quickly we match the exponents and rewrite the equation as follows:

$$\begin{array}{r} 2.76 \times 10^{4} \\ + \; 69.1 \times 10^{4} \\ \hline \end{array}$$

This is similar to the algebraic equation:

$$\begin{array}{r} 2.76y \\ + \; 69.1y \\ \hline \end{array}$$

where $y = 10^4$. We simply add the coefficients of y. Rounding, we have $3y + 69y = 72y$. Thus 72×10^4, or 7.2×10^5 is the answer.

When rearranging 6.91×10^5 to 69.1×10^4, we simply multiply by 10/10 (a form of 1). In other words, we divide 72 by 10 and multiply 10^4 by 10.

$$\overset{\times 10}{6.91} \times \underset{\div 10}{10^5} = 69.1 \times 10^4$$

A useful mnemonic for remembering which way to move the decimal point when we add or subtract from the exponent is to use the acronym LARS,

Left **A**dd, **R**ight **S**ubtract

---

i-7 Multiplication and Division

When multiplying similar bases with exponents add the exponents; when dividing, subtract the exponents. $10^4 \times 10^5 = 10^9$. $10^4/10^{-6} = 10^{10}$.

When multiplying or dividing with scientific notation, we deal with the exponential terms and mantissa separately, *regardless of the number of terms*. For instance:

$$\frac{(3.2\times10^4)\times(4.9\times10^{-8})}{(2.8\times10^{-7})}$$

should be rearranged to:

$$\frac{3\times5}{3}\times\frac{10^4\times10^{-8}}{10^{-7}}$$

giving us an MCAT answer of something greater than 5×10^3. (The exact answer, 5.6×10^3, is greater than our estimate because we decreased one term in the numerator by more than we increased the other, which would result in a low estimate, and because we increased the term in the denominator, which also results in a low estimate.)

When taking a term written in scientific notation to some power (such as squaring or cubing it), we also deal with the decimal and exponent separately. The MCAT answer to:

$$(3.1 \times 10^7)^2$$

is something greater than 9×10^{14}. Recall that when taking an exponential term to a power, we multiply the exponents.

The first step in taking the square root of a term in scientific notation is to make the exponent even. Then we take the square root of the mantissa and exponential term separately.

$$\sqrt{8.1\times10^5}$$

Make the exponent even.

$$\sqrt{81\times10^4}$$

Take the square root of the mantissa and exponential term separately.

$$\sqrt{81} \times \sqrt{10^4} = 9 \times 10^2$$

Notice how much more efficient this method is. What is the square root of 49,000? Most students start thinking about 700, or 70, or something with a 7 in it. By using the scientific notation method, we quickly see that there is no 7 involved at all.

$$\sqrt{49{,}000} \times \sqrt{4.9 \times 10^4} = 2.1 \times 10^2$$

Try finding the square root of 300 and the square root of 200.

Questions:

Solve the following problems without a calculator. Try not to use a pencil.

1. $\dfrac{2.3\times10^{7}\times5.2\times10^{-5}}{4.3\times10^{2}}$

A. 1.2×10^{-1}
B. 2.8
C. 3.1×10
D. 5.6×10^{2}

2. $(2.5 \times 10^{-7} \times 3.7 \times 10^{-6}) + 4.2 \times 10^{2}$

A. 1.3×10^{-11}
B. 5.1×10^{-10}
C. 4.2×10^{2}
D. 1.3×10^{15}

3. $[(1.1 \times 10^{-4}) + (8.9 \times 10^{-5})]^{1/2}$

A. 1.1×10^{-2}
B. 1.4×10^{-2}
C. 1.8×10^{-2}
D. 2.0×10^{-2}

4. $\frac{1}{2}(3.4 \times 10^{2})(2.9 \times 10^{8})^{2}$

A. 1.5×10^{18}
B. 3.1×10^{18}
C. 1.4×10^{19}
D. 3.1×10^{19}

5. $\dfrac{1.6\times10^{-19}\times15}{36^{2}}$

A. 1.9×10^{-21}
B. 2.3×10^{-17}
C. 1.2×10^{-9}
D. 3.2×10^{-9}

Answers:

1. **B is correct.** The exact answer is 2.7814.
2. **C is correct.** The other numbers are insignificant.
3. **B is correct.** The exact answer is 1.4107×10^{-2}.
4. **C is correct.** The exact answer is 1.4297×10^{19}.
5. **A is correct.** The exact answer is 1.8519×10^{-21}.

i-8 Proportions

On the MCAT, proportional relationships between variables can often be used to circumvent lengthy calculations or, in some cases, the MCAT question simply asks the test taker to identify the relationship directly. When the MCAT asks for the change in one variable due to the change in another, they are making the assumption that all other variables remain constant.

In the equation $F = ma$, we see that if we double F while holding m constant, a doubles. If we triple F, a triples. The same relationship holds for m and F. This type of relationship is called a **direct proportion**.

$$\overset{2\nearrow}{F} = m\overset{2\nearrow}{a}$$

F and a are directly proportional.

Notice that if we change the equation to $F = ma + b$, the directly proportional relationships are destroyed. Now if we double F while holding all variables besides a constant, a increases, but does not double. **In order for variables to be directly proportional to each other, they must both be in the numerator or denominator when they are on opposite sides of the equation, or one must be in the numerator while the other is in the denominator when they are on the same side of the equation. In addition, all sums or differences in the equation must be contained in parentheses and multiplied by the rest of the equation. No variables within the sums or differences will be directly proportional to any other variable.**

If we examine the relationship between m and a, in $F = ma$, we see that when F is held constant and m is doubled, a is reduced by a factor of 2. This type of relationship is called an **inverse proportion**. Again the relationship is destroyed if we add b to one side of the equation. **In order for variables to be inversely proportional to each other, they must both be in the numerator or denominator when they are on the same side of the equation, or one must be in the numerator while the other is in the denominator when they are on opposite sides of the equation. In addition, all sums or differences in the equation must be contained in parentheses and multiplied by the rest of the equation. No variables within the sums or differences will be directly proportional to any other variable.**

$$F = \underset{\swarrow 2}{m}\overset{2\nearrow}{a}$$

m and a are inversely proportional.

If we examine a more complicated equation, the same rules apply. However, we have to take care when dealing with exponents. One method to solve an equation using proportions is as follows. Suppose we are given the following equation:

$$Q = \frac{\Delta P \pi r^4}{8\eta L}$$

This is Poiseuille's Law. The volume flow rate Q of a real fluid through a horizontal pipe is equal to the product of the change in pressure ΔP, π, and the radius of the pipe to the fourth power r^4, divided by 8 times the viscosity η and the length L of the pipe.

Water ($\eta = 1.80 \times 10^{-3}$ Pa s) flows through a pipe with a 14.0 cm radius at 2.00 L/s. An engineer wishes to increase the length of the pipe from 10.0 m to 40.0 m without changing the flow rate or the pressure difference. What radius must the pipe have?

A. 12.1 cm
B. 14.0 cm
C. 19.8 cm
D. 28.0 cm

Answer: The only way to answer this question is with proportions. Most of the information is given to distract you. Notice that the difference in pressure between the ends of the pipe is not even given and the flow rate would have to be converted to m^3/s. To answer this question using proportions, multiply L by 4 and r by x. Now pull out the 4 and the x. We know from that, by definition, $Q = \Delta P\pi r^4/8\eta L$; thus, $x^4/4$ must equal 1. Solve for x, and this is the change in the radius. The radius must be increased by a factor of about 1.4. $14 \times 1.4 = 19.6$. The new radius is approximately19.6 cm. The closest answer is C.

$$Q = \frac{\Delta P\pi r^4}{8\eta L}$$

$$Q = \frac{\Delta P\pi (xr)^4}{8\eta (4L)}$$

$$Q = \frac{\Delta P\pi r^4}{8\eta L} \times \frac{x^4}{4}$$

$$4 = x^4$$

$$x = \sqrt{2}$$

Questions:

1. The coefficient of surface tension is given by the equation $\gamma = (F - mg)/(2L)$, where F is the net force necessary to pull a submerged wire of weight mg and length L through the surface of the fluid in question. The force required to remove a submerged wire from water was measured and recorded. If an equal force is required to remove a separate submerged wire with the same mass but twice the length from fluid x, what is the coefficient of surface tension for fluid x. ($\gamma_{water} = 0.073$ mN/m)

A. 0.018 mN/m
B. 0.037 mN/m
C. 0.073 mN/m
D. 0.146 mN/m

2. A solid sphere rotating about a central axis has a moment of inertia

$$I = \frac{2}{3}MR^2$$

where R is the radius of the sphere and M is its mass. Although Callisto, a moon of Jupiter, is approximately the same size as the planet Mercury, Mercury is 3 times as dense. How do their moments of inertia compare?

A. The moment of inertia for Mercury is 9 times greater than for Callisto.
B. The moment of inertia for Mercury is 3 times greater than for Callisto.
C. The moment of inertia for Mercury is equal to the moment of inertia for Callisto.
D. The moment of inertia for Callisto is 3 times greater than for Mercury.

3. The force of gravity on an any object due to earth is given by the equation $F = G(m_oM/r^2)$ where G is the gravitational constant, M is the mass of the earth, m_o is the mass of the object and r is the distance between the center of mass of the earth and the center of mass of the object. If a rocket weighs 3.6×10^6 N at the surface of the earth what is the force on the rocket due to gravity when the rocket has reached an altitude of 1.2×10^4 km? ($G = 6.67 \times 10^{-11}$ Nm²/kg², radius of the earth = 6370 km, mass of the earth = 5.98×10^{24} kg)

A. 1.2×10^5 N
B. 4.3×10^5 N
C. 4.8×10^6 N
D. 9.6×10^6 N

4. The kinetic energy E of an object is given by $E = \frac{1}{2}mv^2$ where m is the object's mass and v is the velocity of the object. If the velocity of an object decreases by a factor of 2 what will happen its kinetic energy?

A. Kinetic energy will increase by a factor of 2.
B. Kinetic energy will increase by a factor of 4.
C. Kinetic energy will decrease by a factor of 2.
D. Kinetic energy will decrease by a factor of 4.

5. Elastic potential energy in a spring is directly proportional to the square of the displacement of one end of the spring from its rest position while the other end remains fixed. If the elastic potential energy in the spring is 100 J when it is compressed to half its rest length, what is its energy when it is compressed to one fourth its rest length.

A. 50 J
B. 150 J
C. 200 J
D. 225 J

Answers:

1. **B is correct.** γ and L are inversely proportional.

2. **B is correct.** Since the bodies are the same size and Mercury is 3 times denser, Mercury is 3 times more massive. Mass is directly proportional to moment of inertia.

3. **B is correct.** If you are good with scientific notation, it is easy to see that r is tripled. r is the distance from the center of the Earth to the earth's surface. The Satellite is two Earth radii from the surface of the Earth, so it is three Earth radii from the center of the earth. Since the square of r is inversely proportional to F, F must be divided by 9.

4. **D is correct.** E is directly proportional to v^2.

5. **D is correct.** If we imagine a spring 100 cm long at rest (We can use any spring length but 100 is always a good choice.) then the initial displacement is 50 cm and the final displacement is 75 cm. The displacement is increased by a factor of 1.5 thus the energy is increased by a factor of 1.5^2. 1.5^2 is greater than 1.4^2 or greater than 2. Thus the energy is greater than 2×100.

i-9 Graphs

The MCAT requires that you recognize the graphical relationship between two variables in certain types of equations. The three graphs below are the most commonly used. You should memorize them. The first is a directly proportional relationship; the second is an exponential relationship; and the third is an inversely proportional relationship.

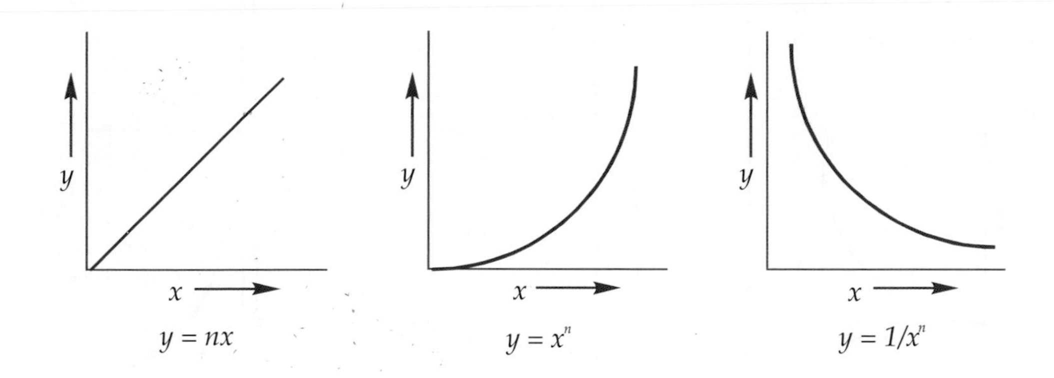

(Note: n is greater than zero for the graph of $y = nx$, and n is greater than one for the other two graphs.)

Notice that, if we add a positive constant b to the right side, the graph is simply raised vertically by an amount b. If we subtract a positive constant b from the right side, the graphs are shifted downwards.

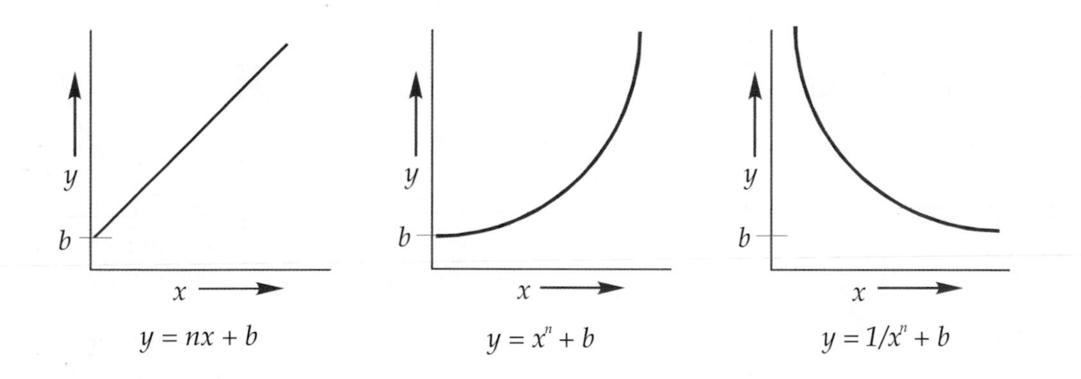

As long as the value of n is within the given parameters, the general shape of the graph will not change. When graphs are unitless, multiplying the right side of an equation by a positive constant will not change the shape of the graph. If one side of the equation is negative, or multiplied by a negative constant, the graph is reflected across the x axis.

Whenever the MCAT asks you to identify the graphical relationship between two variables you should assume that all other variables in the equation are constants unless told otherwise. Next, manipulate the equation into one of the above forms (with or without the added constant b, and choose the corresponding graph.

If you are unsure of a graphical relationship, plug in 1 for all variables except the variables in the question and then plug in 0, 1, and 2 for x and solve for y. Plot your results and look for the general corresponding shape.

Questions:

1. The height of an object dropped from a building in the absence of air resistance is given by the equation $h = h_o + v_o t + \frac{1}{2} gt^2$, where h_o and v_o are the initial height and velocity respectively and g is -10 m/s^2. If v_o is zero which graph best represents the relationship between h and t?

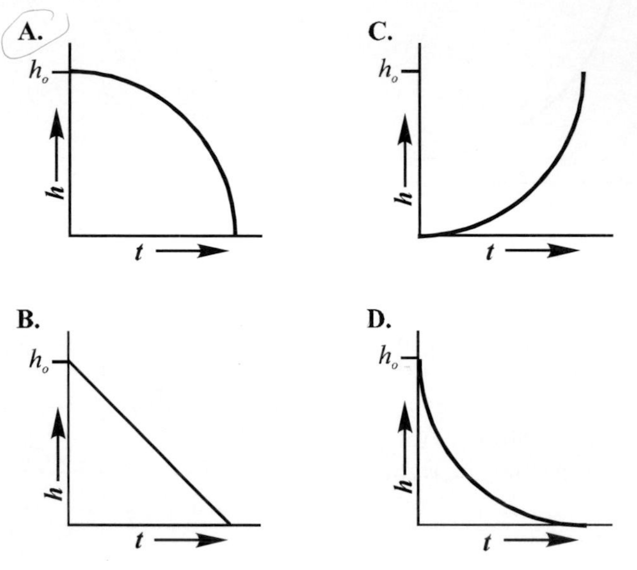

2. Which of the following graphs best describes the magnitude of the force (F) on a spring obeying Hooke's law ($F = -k\Delta x$) as it is compressed to Δx_{max}?

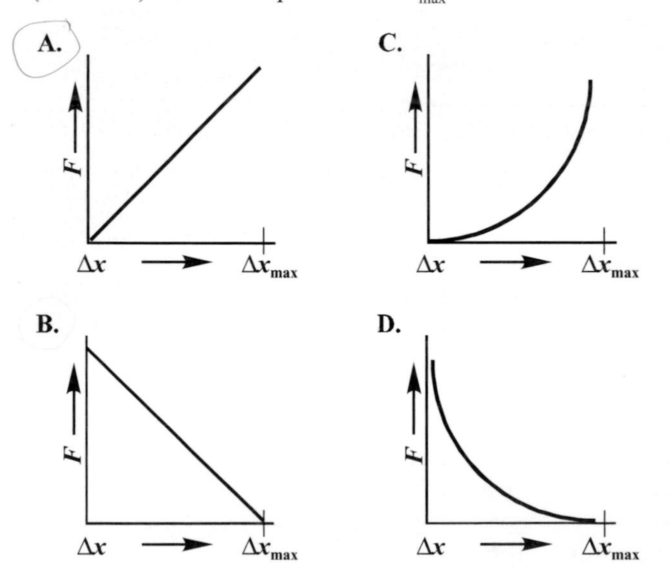

3. Which of the following graphs shows the relationship between frequency and wavelength of electromagnetic radiation through a vacuum? ($c = \nu\lambda$)

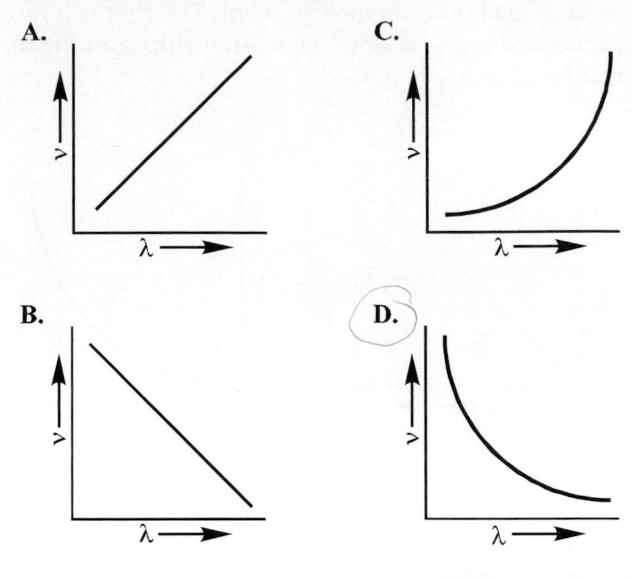

4. Which of the following graphs best describes the magnitude of the electrostatic force $F = k(qq)/r^2$ created by an object with negative charge on an object with a positive charge as the distance r between them changes?

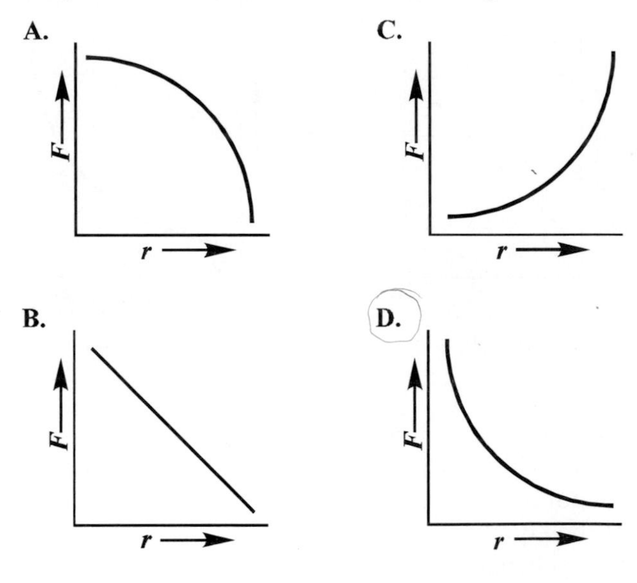

5. Which of the following graphs demonstrates the relationship between power P and work W done by a machine? ($P = W/t$)

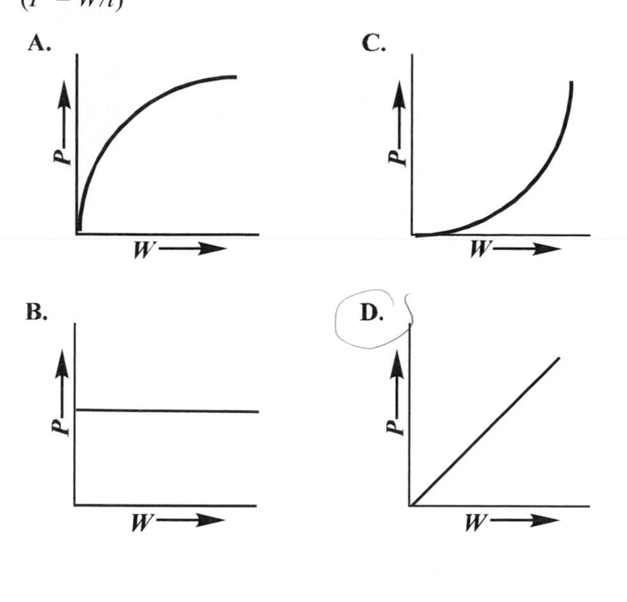

Answers:

1. **A is correct.** Since v_o is zero we have $h = h_o + \frac{1}{2}gt^2$. Since g is in the opposite direction to h, and ho is a constant we can rewrite this equation as $h = -\frac{1}{2}gt^2 + h_o$ where $g = 10$. This is the same form as $y = x^n$. The negative sign flips the graph vertically. In addition a constant has been added to the right side so the graph intercepts the y axis at h_o.
2. **A is correct.** The question asks for magnitude. Thus the negative sign is ignored and the equation has the form $y = nx$.
3. **D is correct.** Manipulation of this formula produces $\nu = c/\lambda$. Which is in the form of $y = 1/x$.
4. **D is correct.** The form of this equation is $y = 1/x^n$. The negative can be ignored because the question asks for magnitude.
5. **D is correct.** The equation has the form $y = nx$ where n is $1/t$.

Lecture 1

Strategy and Tactics

1-1 The Layout of the Verbal Reasoning Section

The Verbal Reasoning Section of the MCAT is composed of nine passages, averaging 600 words per passage. Generally, a passage discusses an area from the humanities, social sciences, or natural sciences. Six to ten multiple-choice questions follow each passage for a total of 60 questions. Answers to these questions do not require information beyond the text of the passage. The test taker has 85 minutes to complete the entire section.

1-2 Other Verbal Strategies

Dogma about the Verbal Section is abundant and free, and that's an accurate reflection of its value. There are many cock-a-mamie verbal strategies touted by various prep companies, academics, and well-wishers. **We strongly suggest that you ignore them.** Some test prep companies design their verbal strategy to be marketable (to make money) as opposed to being efficient (raise your score); the idea being that unique and strange will be easier to sell than commonplace and practical. Desperate techniques such as note taking and skimming are prime examples.

Some colleges offer classes designed specifically to improve reading comprehension in the MCAT Verbal Section. Typically, such classes resemble English 101 and are all but useless at improving your score. They are often taught by well-meaning humanities professors who have never even seen a real MCAT verbal section. Being a humanities professor does not qualify you as an expert at the MCAT Verbal Section. The emphasis in such classes is usually on detailed analysis of what you read rather than how to eliminate wrong answers and find correct answers. Improvements are predictably miserable.

There are those who will tell you that a strong performance on the verbal section requires speed-reading techniques. This is not true. Most speed-reading techniques actually prove to be an impediment to score improvements by shifting focus from comprehension to reading technique. It is unlikely that you will improve both your speed and comprehension in a matter of weeks. As you will soon see, speed is not the key to a good MCAT verbal score. Finishing the Verbal Section is within the grasp of everyone, if they follow the advice posited by this book.

A favorite myth of MCAT students is that copious amounts of reading will improve scores on the Verbal Section. This myth originated years ago when one prep company having insufficient verbal practice materials suggested to their students to "read a lot" rather than use the other companies materials. The myth has perpetuated itself ever since. "Reading a lot" is probably the least efficient method of improving your verbal score. If you intend to take the MCAT four or five years hence, you should begin "reading a lot". If you want to do well on the verbal this year, use the strategies that follow.

1-3 Take Our Advice

Most smart students listen to advice, then pick and choose the suggestions that they find reasonable while disregarding the rest. This is not the most efficient approach for preparing to take the MCAT Verbal Section. In fact, it is quite counter productive. Please abandon all your old ideas about verbal and follow our advice to the letter. Don't listen to your friends and family. They are not experts at teaching students how to score well on the MCAT Verbal Reasoning Section. We are.

1-4 Expected Improvement

Taking the MCAT verbal section is an art. (Not exactly what a science major wants to hear.) Like any art form, improvement comes gradually with lots of practice. Imagine attending a class in portraiture taught by a great artist. You wouldn't expect to become a Raphael after your first lesson, but you would expect to improve after weeks of coaching. The verbal section is the same way. Follow our directions to the letter, and with practice you will see dramatic improvements over time.

1-5 The Examkrackers Approach to MCAT Verbal Reasoning

We shall examine the verbal section on two levels: strategic and tactical. The strategic point of view will encompass the general approach to the section as a whole. The tactical point of view will explain exactly what to do, passage by passage, and question by question.

Strategy

There are four aspects to strategy:

1. **Energy**
2. **Focus**
3. **Confidence**
4. **Timing**

Energy

Pull your chair close to the table. Sit up straight. Place your feet flat on the floor, and be alert. This may seem to be obvious advice to some, but it is rarely followed. Test-takers often look for the most comfortable position to read the passage. Do you really believe that you do your best thinking when you're relaxed? Webster's Dictionary gives the definition of relaxed as "freed from or lacking in precision or stringency." Is this how you want to be on your MCAT? Your cerebral cortex is most active when your sympathetic nervous system is in high gear, so don't deactivate it by relaxing. Your posture makes a difference to your score.

One strategy of the test writers is to wear you down with the verbal section before you begin the biology section. You must mentally prepare yourself for the tremendous amount of energy necessary for a strong performance on the verbal section. Like an intellectual athlete, you must train yourself to concentrate for long periods of time. You must improve your reading comprehension stamina. **Practice! Practice! Practice!** always under timed conditions. **And always give 100% effort when you practice.** If you give less than 100% when you practice, you will be teaching yourself to relax when you take the verbal section, and you will be lowering your score. It is more productive to watch TV than to practice with less than complete effort. If you are not mentally worn after finishing three or more verbal passages, then you have not tried hard enough, and you have trained yourself to do it incorrectly; you have lowered your score. Even when you are only practicing, sit up straight in your chair and attack each passage.

Focus

The verbal section is made up of nine passages with both interesting and boring topics. It is sometimes difficult to switch gears from "the migration patterns of the Alaskan tit-mouse" to "economic theories of the post-Soviet Union." In other words, sometimes you may be reading one passage while thinking about the prior passage. You must learn to **focus your attention on the task at hand.** We will discuss methods to increase your focus when we discuss tactics.

During the real MCAT, it is not unlikely that unexpected interruptions occur. People get physically ill, nervous students breathe heavily, air conditioners break down, and lights go out. Your score will not be adjusted for unwelcome interruptions, and excuses will not get you into med school, so learn to focus and **ignore distractions**.

Confidence

There are two aspects to confidence on the Verbal Section: 1) **be confident of your score** and 2) **be arrogant when you read.**

Imagine taking a multiple choice exam and narrowing 50% of the questions down to just two answer choices, and then guessing. On a physics exam, this would almost certainly indicate a very low grade. Yet, this exact situation describes a stellar performance on the Verbal Section of the MCAT. Everyone of whom we know that has earned a perfect score on the Verbal Section (including many of our own students) has guessed on a large portion of the answers. The test writers are aware that most students can predict their grade on science exams based upon their performance, and that guessing makes science majors extremely uncomfortable. The Verbal Section is the most dissatisfying in terms of perceived performance. You should realize that even the best test takers finish the Verbal Section with some frustration and insecurity concerning their performance. A perceived dissatisfactory performance early in the testing day is likely to reflect poorly in scores on the Biology Section. You should assume that you have guessed correctly on every answer of the verbal section and get psyched to ace the Biological Sciences Section.

The second aspect of confidence concerns how you read the passage. Read the passages as if you were a Harvard professor grading high school essays. Read critically. If you are confused while reading the passage, assume that it is the passage writer, and not you, who is at fault. If you find a contradiction in the reasoning of the argument, trust your reasoning ability that you are correct. The questions will focus on the author's argument and you must be confident of the strong and weak points. In order to identify the strong and weak points, you must read with confidence, even arrogance.

Timing

If you want a 10 or better on the Verbal Section, you must read every passage and attempt to answer every question. If you want to go to medical school, you should attempt to score 10 on the Verbal Section. Therefore, **read every passage in the order given, and attempt every question.**

Skipping around in the Verbal Section to find the easiest passages is a marketable strategy for a prep company but an obvious waste of time for you. It is a bad idea that makes a lot of money for some prep companies because it's an easy trick to sell. 'Cherry picking' is an unfortunate carry over from SAT strategy where it works because the questions are prearranged in order of difficulty. On the MCAT, some passages are difficult to read, but have easy questions; some passages are easy to read, but have difficult questions. Some passages start out difficult and finish easy. You have no way of knowing if a passage is easy or difficult until you have read the entire passage and attempted all the questions, so 'cherry picking' lowers your score.

If you begin reading a passage and are asking yourself "Shall I continue, or shall I move on to the next passage? Is this passage easy or difficult?", then you are not reading with confidence; you are not concentrating on what the author is saying; and you are wasting valuable time. Your energy and focus should be on doing well on each passage, not on trying to decide which passage to do first.

Hmmm. Let's see. I must knock down all nine blocks. Is it faster and more efficient to knock them down in order, or is it faster to decide which one is heaviest and then run back and forth and knock them down out of order?

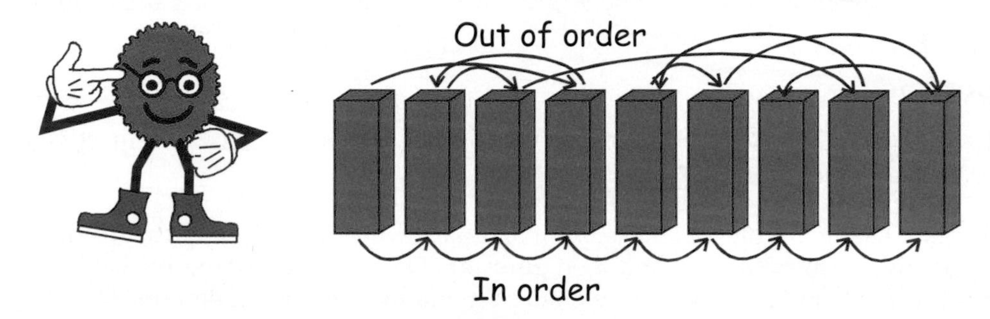

Check your timer only once during the Verbal Section. Constantly checking your timer is distracting and not very useful since some passages take longer to finish than others. Instead, **check your timer only once, and after you have finished the fifth passage.** You should have about 40 minutes left. A well-practiced test taker will develop a sense of timing acute enough to obviate looking at a timer at all.

Don't spend too much time with the difficult questions. **Guess at the difficult questions and move on.** Guessing is very difficult for science students, who are accustomed to being certain of the answer on an exam or getting the answer wrong. Test writers are aware of this, and use it to their advantage. You should learn to give up on difficult questions so that you have more time on easier questions. Accurate guessing on difficult questions is one of the keys to finishing the exam and getting a perfect score. To accurately guess, you must learn to use all your tools for answering the questions. We will discuss this when we discuss tactics.

Many test-takers are able to guess on difficult questions during a practice exam, but when it comes to the real MCAT, they want to be certain of the answers. This meticulous approach has cost such students dearly on their scaled score. Learn to guess at difficult questions so you have time to answer the easy questions.

Finish the entire section with two minutes to spare, no more, no less. If you have more than two minutes to spare, you missed questions on which you could have spent more time. Finishing the exam early and returning to difficult questions is not a good strategy. The stress of exam taking actually makes you more perspicacious while you take the exam. When you finish an exam, even if you intend to go back and check your work, you typically breathe a sigh of relief. Upon doing so, you lose your perspicacity. The best strategy is to use your time efficiently during your first and only pass through the exam.

Some people have difficulty finishing the exam. These people often think that they can't finish because they read too slowly. This is not the case. In tactics, we will discuss how finishing the exam does not depend upon reading speed and that anyone can finish the exam.

1-6 Tactics

Although, at first glance, it may not appear so, the following techniques are designed to increase your pace and efficiency. Tactics is where many students begin to pick and choose a verbal method that they think best suits their own personality. Please don't do this. Follow these steps exactly for each passage and after much practice your verbal score will move to a ten or above.

- **Take a five second break**
- **Read every word**
- **Construct a main idea**
- **Use all four tools to answer the questions:**
  1. going back;
  2. the main idea;
  3. the question stems; and
  4. the answer choices.

The Five Second Break

If you were to observe a room full of MCAT takers just after the sentence "You may break the seal on your test booklet and begin," you would see a room full of people frantically tear open their test booklets, read for 20 to 30 seconds, pause, and then begin rereading from the beginning. Why? Because as they race through the first passage, they are thinking about what is happening to them ("I'm taking the real MCAT! Oh my God!"), and not thinking about what they are reading. They need a moment to become accustomed to the idea that the MCAT has actually begun. They need a moment to focus. However, they don't need 20 to 30 seconds! They take so much time because they are trying to do two things at once; calm themselves down and understand the passage. They end up accomplishing neither. This loss of concentration may also occur at the beginning of each new passage, when the test-taker may still be struggling with thoughts of the previous passage while reading the new passage.

If you continued to observe the test-takers, you would see them in the midst of a passage suddenly stop everything, lift up their head, stretch, yawn, or crack their knuckles. This is their beleaguered mind forcing them to take a break. No one has an attention span 85 minutes long. If you don't allow yourself a break, your mind will take one. How many times have you been reading a passage when suddenly you realize, you weren't concentrating? You're forced to start the passage over. More time is wasted.

There is a simple method to prevent all this lost time. Instead of taking breaks at random, inconvenient moments, plan your breaks. **Before each passage, including the first passage, take five seconds to focus your thoughts.** Remind yourself to forget the last passage and all other thoughts not related to the task at hand. Remind yourself to sit up straight, concentrate, and focus. For these five seconds, look away from the page, stretch your muscles and prepare to give your full attention to the next passage. Then begin and don't break your concentration until you have finished answering all the questions to that passage. The five second break is like a little pep-talk before each passage.

Unfortunately, most students will not take the five second break. Understand one thing. All students will take breaks during the verbal section. Most will take them without realizing it, and most will take them at inopportune moments. If your goal is to get the highest verbal score of which you are capable, you should take the five second break at planned intervals.

Reading the Passage

Most test takers have difficulty finishing the verbal section in the 85 minutes allowed. Many finish as few as six passages. Strangely enough, any premed without a reading disorder is capable of reading 5400 words and taking a one hour nap in 85 minutes. A very slow reader can easily read every word of a 600 word passage in 3 minutes. Try it! It's true! This leaves 58 minutes to answer the questions, or nearly one minute per question to answer the questions. In other words, over two thirds of your time is spent answering questions on the MCAT Verbal Section, and less than one third is spent reading. If you read TWICE as fast as you do now, you would have about 70 seconds, instead of 60 seconds, to answer each question. **So increasing your reading speed has very little effect on your verbal score.** If you're not finishing now, you won't finish by reading faster.

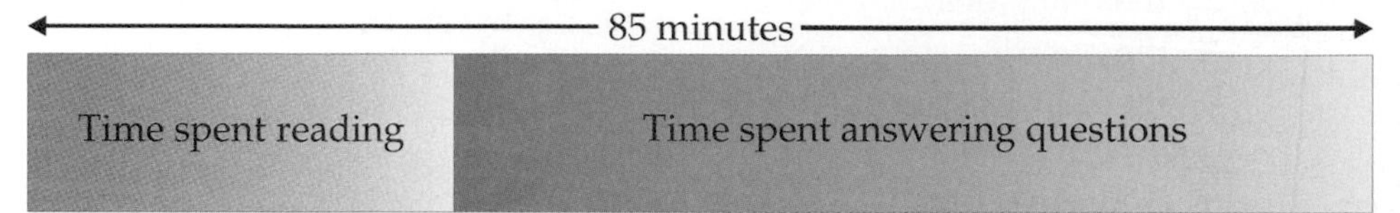

Improving your efficiency at answering questions will be more profitable than increasing your reading speed and allow you more time to read the passage. If you increase your reading speed by 10%, a strong improvement, you will only gain 2 minutes and 12 seconds on the entire exam. Spread over 60 questions, this allows you an additional 2.2 seconds per question. Not too fruitful. If you increase your efficiency at answering questions by 10%, a rather simple task as you will soon see, you gain 5 minutes and 48 seconds. This is almost enough time to read two additional passages!

So why do so many test-takers fail to finish the verbal section? The answer is "because they spend too much time hunting for the answer in the passage, and end up reading the passage many times over." We'll talk more about "going back" to the passage when we discuss where to find the answer choice. For now, just believe us that **you can read every word in the verbal section and easily finish the exam**, so you should.

Have you ever tried skimming through a novel, not reading every word? Try it and see how much you understand. If you don't usually understand much when you skim, then why would you skim on the most important test of your life; especially when doing so won't give you much more time to answer the questions? **Don't skim.**

Have you ever mapped out a novel by writing a brief synopsis alongside each paragraph as you read? Try it. We think you will fall asleep from boredom. You will understand less of what you read, not more. Passages are intended to be read in their entirety as a single work presenting one overriding theme. MCAT expects you to understand this theme. The details within this theme are far less important. **Don't distract yourself by writing in the margins.**

The people that write the MCAT know that most of us are scientists. They know that we like to find the exact answer to things. Give us a mysterious powder and let us analyze it, and we will tell you exactly what it is. Show us the exact words in a passage as an answer choice and we will probably choose it. Don't fall for this trap. The Verbal Section tests your ability to detect and understand ambiguities and gray areas, not details. Rely heavily on your main idea and give little weight to details. If you are highly certain of all your answers on the Verbal Section, then you probably have fallen for all its traps. **Mastering this section is as much an art as a science.** With practice, you will develop a 'feel' for a good MCAT answer. Learn to use this 'feel' to help you move faster through the Verbal Section. If you teach yourself not to expect the concrete certainty that you get with science questions, you will become more comfortable with the Verbal Section and your score will increase.

The biggest mistake you can make on the verbal section is to consciously attempt to remember what you are reading. The vast majority of the questions will not concern the details of the passage and will not be answerable by searching the passage for facts. Most questions are about the main idea of the passage. The main idea will not be found in a list of details. In order to learn the main idea, the passage as a whole must be understood. Read the passage the way you would read an interesting novel; **concentrate on the main idea, not the detail.**

When I create a great soup, you do not taste the salt, and each spice separately. You must experience the whole soup as a single, wonderful consomme'. Otherwise you are nothing but a tasteless peasant, and I will not invite you to dinner.

An often posited tactic is to read the questions first; don't do it! You will not remember even one question while you read the passage, much less the 6 to 10 questions that accompany every passage. In fact, a short term memory can contain 5 items; that may be why the passages are followed by six or more questions. Not only that, reading the questions first will force you to read for detail and you will never learn the main idea. You will probably end up rereading the passage many times but never straight through. This results in a tremendous waste of time.

Don't circle or underline words. This is another marketing technique that has sold well but is counterproductive. It is very unlikely that underlining or circling a sentence or a word will assist you in answering any questions. Have you ever answered a question correctly because you underlined or circled something in the passage? Underlining and circling words forces you to concentrate on detail; fine for the SAT, not good for the MCAT. When you underline or circle something, you are reading it at least twice. This interrupts the flow of the passage. It distracts you from the main idea. This is an old SAT trick, inappropriately applied to MCAT.

Some of the Verbal topics will fascinate you and some will bore you. The challenge will be to forget the ones that fascinate you as soon as you move to the next passage, and to pay close attention to the ones that bore you. **Train yourself to become excited and interested in any and every passage topic.** This will increase your comprehension. However, don't become so engrossed in a passage that you slow your pace.

Don't use fancy speed reading techniques where you search for meaningful words or try to read entire phrases in one thought. This will only distract you from your goal. Read the way you normally read. Your reading speed is unlikely to change significantly in 10 weeks, and your reading speed is not the problem anyway. Finishing the entire section depends upon how long you spend on the questions, not how long it takes you to read the passages. You also cannot assume that the passages are written well enough so that you can read just the first and last sentence of each paragraph. They are sometimes barely intelligible when you read every word. You must read every word, read quickly and concentrate.

Read each passage like you are listening to a friend tell you a very interesting story. Allow the details (names, dates, times) to slip in one ear and out the other, while you wait with baited breath for the main point. The funny thing about this type of reading is that, when you practice it, you can't help but remember most of the details. Even if you were to forget some of the details, it only takes about 5 seconds to find a name, number, or key word in a 600 word passage. Thus, when you run into a rare question about a detail that you've forgotten, it is easy to find the answer. Another convenient aspect of this type of reading is that you are trying to accomplish exactly what the verbal section will be testing: your ability to pick out the main idea. The best thing about this type of reading is that you have practiced it every day of your life. This is the way that you read novels, newspapers and magazines. Read the passages the way that you read best; read for the main idea.

When you read, ask yourself, "What is the author trying to say? What is his point? Is he in favor of idea A or against it? If this author were sitting right in front of me, would he want to discuss idea A or is his real interest in idea B?" **Creating an image of the author in your mind will help you understand him.** Use your life experiences to stereotype the author. This will help you make quick, intuitive decisions about how the author might answer each MCAT question about his passage. Make careful mental note of anything the author says that may not fit your stereotype. Use the stereotype to help guide your intuition on the questions.

The Main Idea

When you have finished reading a passage, take twenty seconds and construct a main idea in the form of one or two complete sentences. Verbal Reasoning Lecture 3 will cover how to construct a main idea. On a timed MCAT, writing the main idea requires too much time, so you should **spend 20 seconds mentally contemplating the main idea before you begin the questions.** After you have completed an entire timed exam, scored yourself, and taken a break, it is a good idea to go back to each passage and write the main idea for practice.

Answering the Questions

Answering the questions will be covered thoroughly in Verbal Reasoning Lecture 2. For now, attempt to answer the questions based upon the main idea and not the details.

Answering the Questions

2-1 Tools to Find the Answer

For most students, the Verbal Reasoning Section is literally a test of their ability to comprehend what they have read. Such students read a question and choose the correct answer based upon what was said in the passage. If they do not arrive at an answer, they eliminate answer choices based upon what was said in the passage. If they still don't arrive at an answer, they search the passage for relevant information that they may have missed or don't recall. If they still don't arrive at a single answer choice, which is likely to be about 50% of the time with this method, they repeat the process until they give up and make a random guess in frustration. This method uses only about 50% of the information provided by the test. When you consider that a portion of the questions on a multiple choice test will be answered correctly by luck, it's no coincidence that the national mean score on the MCAT is attainable by answering only about 61% of the questions correctly. When you can't identify an answer, 'thinking harder' (whatever that means) is not an effective solution. Nor is an effective solution to search the passage until the answer jumps out at you. However, both use up your precious time.

In addition to your understanding of the passage, there are four tools that you should use to help you answer the questions. These four tools go beyond your understanding of the passage. They force you to consider additional information presented to you in the question stems and answer choices that is often overlooked or otherwise noticed only on a subconscious level.

The four tools are:

1. going back;
2. the main idea;
3. the question stems; and
4. the answer choices.

Going Back

'Going back' refers to actually rereading parts of the passage to search for an answer. 'Going back' should be **used only when:**

1. you are regularly finishing an exam on time;
2. you know what you're looking for;

   and
3. you know where you can find the answer.

'Going back' is the most time consuming and least useful of the four tools. Unfortunately, it is the tool most often relied upon by inexperienced test takers. It is true that forgotten details can be found by rereading parts of the passage. However, most questions require an understanding of the main idea, not your memory of details. The main idea cannot be found by rereading parts of the passage.

'Word-for-word' and other traps have been set for the unwary test taker looking for the 'feel-good' answer. The 'feel-good' answer is an answer where a section of the passage seems to unequivocally answer the question so that the test taker *feels good* when choosing it. This is often a trap. Remember, the Verbal Section is ambiguous and a simple clear answer is rarely the correct answer.

You should learn to use 'going back' as seldom as possible. Most of the time, you should force yourself to choose the best answer without going back to the passage. This is a difficult lesson to accept, but it is extremely important to achieving your top score. Going back to the passage for anything but a detail will take large amounts of your testing time, and allow the test writers to skew your concept of the main idea by directing you toward specific parts of the passage. **If you are unable to finish the test in the time given, it is because you are overusing the 'going back' tool.** If you are not finishing, you should not go back at all until you can regularly finish.

Questions sometimes refer to line numbers in the passage. **Don't assume that you must go back to the given line number.** Usually these types of questions should be answered without going back to the given line numbers. Often times the most helpful part of the passage in answering the question is nowhere near the lines mentioned. If you do go back, you may have to begin reading 5 lines or more above the actual reference in order to place the reference in the correct context.

Your number one goal should be to finish the Verbal Section. Difficult questions are worth no more than easy questions. **Don't sacrifice five easy questions by spending a long time answering a single difficult question. If you usually finish the Verbal section with time to spare, you can 'go back' to the passage more often; if you don't usually finish the Verbal section, you should stop going back to the passage until you begin finishing within the allotted time on a regular basis.**

"Going back" is a useful tool. Just use it wisely.

Main Idea

The main idea is the most powerful tool for answering MCAT verbal questions. We will discuss the main idea in Verbal Reasoning Lecture 3.

Question Stems

**The section that follows includes material from the MCAT Practice Test 1/Practice Items. These materials are reprinted with permission of the Association of American Medical Colleges (AAMC).

The **question stems hold as much information as the passage.** Read them and see how much you can learn about the passage from just the question stems.

1. The author of the passage believes that the fiction written by the current generation of authors:
2. The overall point made by the passage's comparison of movies to fiction is that:
3. According to the passage, John Gardner concedes that preliminary good advice to a beginning writer might be, "Write as if you were a movie camera." The word *concedes* here suggests that:

4. The fact that the author rereads *Under the Volcano* because it has been made into a movie is ironic because it:
5. The passage suggests that a reader who is not bored by a line-by-line description of a room most likely:
6. The passage suggests that if a contemporary writer were to write a novel of great forcefulness, this novel would most likely:
7. The passage places the blame for contemporary writers' loss of readers on the:

Ask yourself some questions about the author. What does he/she do for a living? How does he/she dress? What does he/she like to eat? How does he/she vote? How old is he/she? Is he/she a he or she?

Look closely at each question stem and see what kind of information you get from it. Why are certain adjectives used? Who is John Gardner? What can I learn about the passage from these question stems?

Now, in the space below, write down everything that you can think of that is revealed about the passage from each stem. Include an answer to each of the seven question stems.

(**Warning:** If you read on without writing the answers, you will miss an important opportunity to improve your verbal skills. Once you read on, the effect of the exercise will be ruined.)

1. ______________________________

____

2. ______________________________

____

3. ______________________________

____

4. ______________________________

____

5. ______________________________

____

6. ______________________________

____

7. ______________________________

____

Information that can be gained from the seven previous question stems:

1. From the first question stem, we immediately know that the passage was about the writing of fiction. The word 'current' suggests a comparison between authors of fiction from the past and the present.

2. From the second question stem we learn that there is also a comparison between movies and fiction. We also know that this was central to the author's point. Movies are a 'current' phenomenon. Hmmm. What is the significance of this?

3. In question stem three, you need to wonder "Who is John Gardner?" You know he is not the author of the passage because on the MCAT, you never know the name of the author. Thus, a named identity will be someone whom the author used either to support his point or as an example of someone who has a bad idea. You should decide which. Now, even if the question didn't ask this, you should have asked yourself about the word 'concedes'. When you concede, you give in. So 'concedes' here indicates that Mr. Gardner is giving in to a point when he says "Write as if you were a movie camera." Mr. Gardner's argument must be that writing (or fiction) is not good when it's like the movies, but it is okay to write like a movie camera when you are a beginning writer. Notice how hesitant the wording is. Beginning is stressed by the use of both words 'preliminary' and 'beginning', and the word 'might' is also used.

At this point, you should begin forming a feeling of what this passage was about: movies versus fiction; current fiction versus past fiction; someone implying that movies don't make for good fiction. The author believes something about current fiction and makes a point about fiction and movies. Three question stems with no passage and not even answer choices to the questions, and we can already get a sense of the passage. The remaining question stems will confirm what the passage is about.

4. The fourth question stem indicates that a movie makes the author read a book. The question states that this is ironic. That means the actual result is incongruous with the expected result. Apparently, according to the author's argument, watching a movie should not make him read the book. Thus, part of the author's argument in the passage must be that movies make people less interested in reading. It is also reasonable to assume from this that the author used John Gardner in stem #3 to support his argument, so the author probably believes that fiction written like a movie is not good fiction. Extrapolating further from the comparison of movies to fiction and the stated dichotomy between current and past fiction, the author is probably arguing that current fiction is not as good as old fiction.

5. The fifth passage compares the phrase 'line-by-line description' with the idea of boredom. It is a simple logical jump to equate 'line-by-line description' with past fiction as opposed to current fiction or movies. From our conclusions thus far about the author's argument, it would be logical to conclude that someone who is NOT bored by 'line-by-line descriptions' would NOT be bored by past fiction, but would, in fact, appreciate it as the author obviously does.

6. Question stem six reinforces our conclusion about the author's argument. The 'If' indicates that 'contemporary writers' do not 'write novels of great forcefulness'. Instead, they must be writing novels that resemble movies. The only question is 'What would a novel of great forcefulness' do? Answering this question is as simple as seating the author in front of you and asking him. The amazing thing is that we already have a

stereotypical idea of this author just by reading six question stems! This guy is a college English professor fed up with the quick fix satisfaction offered by movies. He would love a novel of great forcefulness. Does he think that we would appreciate it? Be careful here. He appreciates the novel because he truly believes that the novel itself is great, and not because he thinks he is great or better than everyone else. The answer is yes, he thinks that we would appreciate a novel of great forcefulness as well.

7. This last question stem answers the previous question. The seventh question stem says that current fiction is losing readers. It asks for the explanation. Of course, the author's whole point is to explain why current fiction is losing readership. It is because it is like movies and not forceful like past fiction.

What should be revealing and even shocking to you is that we can accurately answer every question on this actual AAMC passage without reading the passage. In fact, we can accurately answer every question without reading the passage OR the answer choices. Did you realize that there was this much information in the question stems alone? Have you been using this information to answer the questions on the MCAT? If you haven't, you are capable of scoring many points higher on the MCAT Verbal Section. You can't expect to always be able to answer questions without the passage or the answer choices, but you can expect to gain much information about the passage from the question stems.

Compare your answers with the actual answer choices below.

**The section that follows includes material from the MCAT Practice Test 1/Practice Items. These materials are reprinted with permission of the Association of American Medical Colleges (AAMC).

1.

A. lacks the significance of fiction written by previous generations.
B. is, as a whole, no better and no worse than fiction written by previous generations.
C. brilliantly meets the particular needs of contemporary readers.
D. is written by authors who show great confidence in their roles as writers.

2.

A. contemporary authors have strengthened their fiction by the application of cinematic techniques.
B. the film of *Under the Volcano* is bound to be more popular than the novel.
C. great fiction provides a richness of language and feeling that is difficult to re-create in film.
D. contemporary authors would be well advised to become screenwriters.

3.

I. Gardner's approach to writing has been influenced by the competing medium of film.
II. Gardner must have written screenplays at one point in his life.
III. Gardner dislikes the medium of film.

A. I only
B. II only
C. I and II only
D. II and III only

4.

I. seems to go against the overall point of the passage concerning fiction and film.
II. implies that the film version was a box-office failure.
III. hints that the author was dissatisfied with the novel.

A. I only
B. II only
C. III only
D. II and III only

5.

A. prefers the quick fix of the movies.
B. would be bored by a single shot of a room in a film.
C. has no tolerance for movies.
D. displays the attitude demanded by good fiction.

6.

I. confuse and anger lovers of great literature.
II. exist in stark contrast to the typical contemporary novel.
III. win back some of the readers contemporary writers have lost.

A. I only
B. II only
C. I and II only
D. II and III only

7.

I. competition presented by movies.
II. writers themselves.
III. ignorance of the public.

A. I only
B. II only
C. I and II only
D. I, II and III

Answers to the Questions

Question 1

Choice A, the answer to question #1 is exactly what we expected: past fiction is better than current fiction. Notice that we can simplify the choices to:

A. Current fiction is not as good as past fiction.
B. Current fiction is equal to past fiction.
C. Current fiction is good.
D. Current fiction is good.

Simplifying the question and the answer choices can make the correct answer easier to find. We'll discuss simplification later in this lecture. The main idea is all you need to answer this question.

Question 2

Choice C, the answer to question #2 is also exactly what we expected. The choices can be rephrased to:

A. Movies have been good for fiction.
B. Movies are more likeable than fiction.
C. Movies aren't as good as good fiction.
D. Authors of fiction should make movies.

When we put these questions to our author, the choice is obvious.

Question 3

The answers to question #3 are not what we expected. We expected a more sophisticated question pertaining to the use of the word 'concedes'. Although the question told us much about the passage, the answer choices match a much simpler question than we anticipated, "Who is John Gardner?" The choices can be rephrased as:

I. John Gardner has been influenced by movies.
II. John Gardner wrote movies.
III. John Gardner dislikes movies.

Clearly John Gardner has been influenced by movies if he is suggesting that writing like a movie might be good advice for a beginning writer. From the answer choices, we can see that if I is correct, then III is likely to be incorrect. If Gardner dislikes movies, it would be unlikely that he would be influenced by them. II is incorrect because Gardner would not have to have written movies in order to be influenced by them. Even if III were an option, and even if Gardner is like the author, liking good fiction more than movies isn't the same as disliking movies. Choice A is correct.

Question 4

The answer to question #4 is exactly what we expected. The choices can be rephrased to:

I. Seeing the movie shouldn't have made the author read the book.
II. The movie flopped.
III. The author didn't like the book.

Only choice I addresses the 'irony' suggested in the question. Only choice I pertains to the main idea. The answer is A.

Question 5

Choice D, the answer to question #5 is exactly what we expected. The choices can be rephrased to:

A. If you're patient, you'll prefer the fast pace of movies.
B. If you're patient, you won't like waiting for action.
C. If you're patient, you won't have the patience for the fast pace of movies.
D. If you're patient, you'll like the careful pace of good fiction.

Choices A, B, and C seem to be self contradictory.

Question 6

Choice D, the answer to question #6 is exactly what we expected. Remember that 'a novel of great forcefulness' describes past fiction to our author, and our author would expect us to like past fiction. This describes choices II and III.

Question 7

Choice C, the answer to question #7 is exactly what we expected. Choice I restates the main idea that movies have hurt fiction. Certainly, our author is criticizing current authors, so choice II is also true. Choice III is not true based upon our idea that the author would expect us to like a forceful novel. The answer here is C.

2-2 Answer Choices

Each MCAT question has four possible answer choices. One of these will be the correct answer and the other three we will call *distracters*. Typically, when a verbal question is written, the correct answer choice is written first and then distracters are created. Because the correct answer is written to answer a specific question and a distracter is written to confuse, the two can often be distinguished without even referencing the question. In other words, with practice, a good test-taker can sometimes distinguish the correct answer among the distracters without even reading the question or the passage. This is a difficult skill to acquire and is gained only through sufficient practice.

Begin by learning to recognize typical distracter types. Among other things, effective distracters may be: a statement that displays a subtle misunderstanding of the main idea; a statement that uses the same or similar words as in the passage but is taken out of context; a true statement that does not answer the question; a statement that answers more than the question asks; a statement that relies upon information commonly considered true but not given in the passage.

In order to help you recognize distracters, we have artificially created five categories of **suspected distracters.** It is unlikely, but not impossible that the correct answer choice might also fall into one of these categories. Thus, you must use this tool as a guide to assist you in finding the correct answer, and not as an absolute test.

- **Round-About:** a distracter that moves around the question but does not directly answer it
- **Beyond:** a distracter whose validity relies upon information not supplied by (or information *beyond*) the passage
- **Contrary:** a distracter that is contrary to the main idea
- **Simpleton:** a distracter that is very simple and/or easily verifiable from the passage
- **Unintelligible:** a distracter that you don't understand

The Round-About

Round-about distracters simply don't answer the question as asked. They may be true statements. They may even concur with the passage, but they just don't offer a direct answer to the question. A Round-about is the answer you expect from a politician on a Sunday morning talk show; a lot of convincing words are spoken but nothing is really said.

Beyonds

Often times, a distracter will supply information beyond that given in the question and passage without substantiating its veracity. These distracters are called *beyonds*. When you read a beyond, you typically find yourself wondering something like "This answer sounds good, but this passage was on the economics of the post Soviet Union, I don't remember anything about the Russian revolution."

Beyonds can also play upon current events. A passage on AIDS may have a question with an answer choice about cloning. Cloning may be a hot topic in the news, but if it wasn't mentioned in the passage or in the question, you should be very suspicious of it being in an answer choice.

Don't confuse a *beyond* with an answer choice that directly asks you to assume information as true.

Contraries

A *contrary* distracter contradicts the main idea. If the question is not an EXCEPT, NOT or LEAST, the answer choice is extremely unlikely to contradict the main idea. Most answer choices support the main idea in one form or another.

Simpletons

If the correct answers on the Verbal Section were simple, direct, and straight forward, then everyone would do well. Instead, the correct answers are vague, ambiguous, and sometimes debatable. This means that an answer choice that is easily verifiable from a reading of the passage is highly suspect and often incorrect. These answer choices are called *simpletons*. Simpletons are not always the wrong answer choice, but you should be highly suspicious when you see one.

Typical of simpletons is extreme wording like *always* and *never*.

Here's a manufactured example of a simpleton:

> **13.** In mid-afternoon in December in Montana, the author believes that the color of the sky most closely resembles:
>
> **B.** cotton balls floating on a blue sea.

If this were the answer, everyone would choose it. This is unlikely to be the correct answer.

Unintelligibles

Unintelligibles are answer choices that you don't understand. Whether it's a vocabulary word, or a concept, avoid answer choices that you don't understand. These are likely to be traps. Strangely enough, many test takers are likely to choose an answer that confuses them. This is apparently because the MCAT is a difficult test so students expect to be confused. Test writers sometimes purposely use distracters with obscure vocabulary or incomprehensible diction in order to appeal to the test taker who finds comfort in being confused. As a general rule, don't choose an answer that you don't understand unless you can positively eliminate all other choices. Be confident, not confused.

2-3 Identifying the Correct Answer

Besides identifying distracters, you should become familiar with the look and feel of a typical correct answer choice.

Typical correct answer choices contain *softeners*. Softeners are words that make the answer true under more circumstances, such as *most likely, seemed, had a tendency to,* etc. An answer choice with a softener is not necessarily correct; it is just more likely to be correct.

2-4 Simplification of the Question and Answer Choices

It is often helpful to simplify the question and answer choices in terms of the main idea. For instance, reexamining the questions and answer choices from our original seven AAMC question stems we have a passage with the following main idea:

"Great fiction provides a richness of language and feeling that is difficult to recreate in film. Contemporary authors emulating film have lost this richness and their audience with it."

This is a nice complete main idea but can be difficult to understand all at once. It is helpful to simplify it as follows: There is past fiction, current fiction, and movies.

- Past fiction is good;
- Current fiction is bad;
- Current fiction is like movies.

When analyzing the questions and answer choices, restate them in terms of these ideas, keeping in mind that this is a simplification. For instance, a reference to 'a great, forceful novel' or 'a line-by-line description' can be replaced by 'past fiction'. 'The passage suggests' can be replaced by 'the author thinks'. This is much like using the concept of an ideal gas to approximate the behavior of a real gas and then adding the characteristics of a real gas for the detailed work.

Compare the following restatements with the original seven AAMC questions:

Restatement

1. The author believes current fiction is:

   A. not as good as past fiction.
   B. equal to past fiction.
   C. good.
   D. good.

2. The author compares movies to fiction in order to show that:

   A. movies have been good for fiction.
   B. movies are more likeable than fiction.
   C. movies aren't as good as good fiction.
   D. authors of fiction should make movies.

3. John Gardner says, "Write like the movies," therefore:

   I. he has been influenced by movies.
   II. he wrote movies.
   III. he dislikes movies.

4. The author sees a movie that causes him to read a book, this:

   I. weakens his argument.
   II. means the movie was bad.
   III. means the author didn't like the book.

Original Question

1. The author of the passage believes that the fiction written by the current generation of authors:

   A. lacks the significance of fiction written by previous generations.
   B. is, as a whole, no better and no worse than fiction written by previous generations.
   C. brilliantly meets the particular needs of contemporary readers.
   D. is written by authors who show great confidence in their roles as writers.

2. The overall point made by the passage's comparison of movies to fiction is that:

   A. contemporary authors have strengthened their fiction by the application of cinematic techniques.
   B. the film of *Under the Volcano* is bound to be more popular than the novel.
   C. great fiction provides a richness of language and feeling that is difficult to re-create in film.
   D. contemporary authors would be well advised to become screenwriters.

3. According to the passage, John Gardner concedes that preliminary good advice to a beginning writer might be, "Write as if you were a movie camera." The word *concedes* here suggests that:

   I. Gardner's approach to writing has been influenced by the competing medium of film.
   II. Gardner must have written screenplays at one point in his life.
   III. Gardner dislikes the medium of film.

4. The fact that the author rereads *Under the Volcano* because it has been made into a movie is ironic because it:

   I. seems to go against the overall point of the passage concerning fiction and film.
   II. implies that the film version was a box-office failure.
   III. hints that the author was dissatisfied with the novel.

5. The author says that if you like past fiction:

A. you'll like movies.
B. you'll be bored by past fiction.
C. you won't like movies.
D. you'll like past fiction.

6. If a new novel were like old fiction:

I. people who like old fiction wouldn't like the novel.
II. the novel would not be like current fiction.
III. people would like to read it.

7. No one reads current fiction because:

I. movies are as good.
II. current fiction writers write bad fiction.
III. people are ignorant.

5. The passage suggests that a reader who is not bored by a line-by-line description of a room most likely:

A. prefers the quick fix of the movies.
B. would be bored by a single shot of a room in a film.
C. has no tolerance for movies.
D. displays the attitude demanded by good fiction.

6. The passage suggests that if a contemporary writer were to write a novel of great forcefulness, this novel would most likely:

I. confuse and anger lovers of great literature.
II. exist in stark contrast to the typical contemporary novel.
III. win back some of the readers contemporary writers have lost.

7. The passage places the blame for contemporary writers' loss of readers on the:

I. competition presented by movies.
II. writers themselves.
III. ignorance of the public.

2-5 Summary

You have four tools for finding the correct answer (going back, main idea, question stems, and answer choices). In order to get your best MCAT score, you should use all of them. Your fourth tool is the most difficult to master. When evaluating the answer choices for distracters, keep in mind that there are no absolutes, just suspects. When necessary, restate complicated questions using the simplified concepts from the main idea.

STOP!

(DO NOT LOOK AT THE FOLLOWING QUESTIONS UNTIL CLASS. IF YOU WILL NOT BE ATTENDING CLASS, GIVE YOURSELF 30 MINUTES TO COMPLETE THE FOLLOWING SET OF QUESTIONS.)

The following questions come from three passages. Each page represents a different passage. The passages have been removed to force you to pay attention to the questions and the answer choices. Try to answer the questions, and compare your scaled score to your normal practice MCAT score.

Passage 1 (Questions 1–7)

1. According to the passage, an image is a versatile tool that:

A. is always visual, never abstract.
B. can be either abstract or visual.
C. is always abstract, never visual.
D. is neither visual nor abstract.

2. An experiment found that dogs can remember a new signal for only five minutes, whereas six-year-old children can remember the same signal much longer. Based on the information in the passage, this finding is probably explained by the fact that:

A. a human being possesses a larger store of symbolic images than a dog possesses.
B. the human brain evolved more quickly than the brain of a dog.
C. the children were probably much older than the dogs.
D. most dogs are color-blind.

3. In order to defend poets from the charge that they were liars, Sidney noted that "a maker must imagine things that are not" (line 38). Sidney's point is that:

A. a true poet must possess a powerful imagination.
B. in order to create something, one must first imagine.
C. poets are the most creative people in our society.
D. imagination is not a gift unique to poets, but is possessed by all creative people.

4. In the context of the passage, the statement "if thereby we die a thousand deaths, that is the price we pay for living a thousand lives" (lines 52—54) is most likely meant to suggest that:

A. we must guard against using our imaginations toward destructive ends.
B. although imagination sometimes causes pain, its positive aspects outweigh its negative ones.
C. it is possible to be too imaginative for one's own good.
D. without imagination, the uniquely human awareness of death would not exist.

5. Which of the following findings would most weaken the claim that the use of symbolic imagery is unique to humans?

A. Chimpanzees are capable of learning at least some sign language.
B. Certain species of birds are able to migrate great distances by instinct alone.
C. Human beings have larger frontal lobes than do other animals.
D. Some animals have brains that are larger than human brains.

6. It has been said that language does not merely describe reality but actually helps to bring reality into existence. Which of the points made in the passage would best support this claim?

A. To imagine means to make images and move them about in one's head.
B. The tool that puts the human mind ahead of the animal's is imagery.
C. There is no specific center for language in the brain of any animal except the human being.
D. Images play out events that are not present, thereby guarding the past and creating the future.

7. According to the author, the most important images are:

A. words.
B. poetic images.
C. images of the past.
D. images of the future.

Passage 2 (Questions 8–16)

8. Why is the San Luis Valley site being investigated urgently?

A. Artifacts are few in number.
B. Artifacts are being eroded by the wind.
C. Bison bones are few in number.
D. Excessive rainfall is damaging the site.

9. According to the passage, which of the following activities was common to each band of Folsom Indians?

A. Cultivating a number of different crops
B. Eating a wide variety of wild game
C. Interacting with other bands
D. Making tools out of nearby rocks

10. The passage suggests that the presence of human remains, tools, and animal bones at a single location means that:

A. bison and other animals migrated from one place to another.
B. communal tasks were performed at the site.
C. erosion has not yet occurred at the site.
D. extensive interactions occurred among bands of Paleoindians.

11. Assume that a new Folsom hunter site has just been discovered in northern Texas. On the basis of the information contained in the passage, this site would most likely contain all of the following EXCEPT:

A. clusters of bones and tools.
B. human bones.
C. remains of hearths.
D. tools made of Colorado flint.

12. If a Folsom hunter site containing tools made of petrified wood were discovered in Iowa, where there is little petrified wood, this discovery would weaken which of the following conclusions made in the passage?

I. Paleoindians hunted bison.
II. Folsom hunters did not travel great distances.
III. There was little trading among bands of Folsom hunters.

A. I only
B. III only
C. I and II only
D. II and III only

13. According to the passage, bands of Paleoindians did not trade between one another. What is the evidence for this statement?

A. Tools of a band came only from local sources.
B. Tool shapes were unique to each band.
C. Food sources were unique to each band.
D. Each band had its unique language and customs.

14. Given the information contained in the passage, if a large number of deer bones were discovered at the San Luis Valley site, the most likely explanation for their presence would be that the deer:

A. accidentally died at the scene.
B. competed with bison for food.
C. migrated from another region.
D. served as food for the Indians.

15. Which of the following discoveries would most strengthen the hypothesis that Folsom hunters killed the bison they ate?

A. Bone breaks consistent with the shapes of the Folsom hunters' pointed tools
B. No evidence of an alternative animal food source
C. Bison bones at a Folsom site
D. Similar accumulation of bison bones at many Folsom sites

16. If the Paleoindians had eaten small game such as rabbits instead of large game, the finding of small animal skeletons and individual tools with many edges at the same sites would LEAST support the conclusion that:

A. certain tools had many uses.
B. small animals made up the people's main diet.
C. the animals were killed at the site.
D. tools were used to prepare the animals for use.

Passage 3 (Questions 17–23)

17. The example concerning Galileo (lines 23–31) best supports the author's claim that:

A. science and society usually coexist harmoniously.
B. science works in an unpredictable manner.
C. cultural bias limits scientific theorizing.
D. scientific fact occasionally forces a change in cultural assumptions.

18. Based on the passage, a scientific claim has the best chance of being free from cultural influence when the claim has:

A. much supporting evidence and much social impact.
B. little supporting evidence and little social impact.
C. much supporting evidence and little social impact.
D. little supporting evidence and much social impact.

19. The author mentions the abandonment of eugenics in America and Hitler's use of arguments for sterilization and racial purification primarily to support the claim that:

A. science is often misused.
B. science is impartial.
C. scientific attitudes are sometimes affected by social movements.
D. science should avoid involvement in social issues.

20. The author believes that the view that science is an "inexorable march toward truth" (lines 66—67) is:

A. one of the myths of science.
B. supported by good evidence.
C. clearly proven by the case of Galileo.
D. accepted by most historians of science.

21. According to the author, most historians of science do NOT believe that:

A. scientific facts lead to effective theories.
B. most theories are developed by straightforward induction from facts.
C. objectivity is a worthwhile goal in scientific investigation.
D. facts are influenced by cultural assumptions.

22. When the author states that "science cannot escape its curious dialectic" (line 54), he is emphasizing science's:

A. dilemma between truth and mere theories.
B. interrelationship with social factors.
C. quest for truth.
D. imprecise methodology.

23. According to the author, one reason that scientists have a difficult time escaping cultural assumptions is that scientists often:

A. formulate hypotheses that can only result in the verification of accepted beliefs.
B. project their research findings onto society.
C. attribute too much significance to scientific data as opposed to social belief.
D. base theories on too much data.

STOP. IF YOU FINISH BEFORE TIME IS CALLED, CHECK YOUR WORK. YOU MAY GO BACK TO ANY QUESTION IN THIS TEST BOOKLET.

STOP.

Don't look at the answers yet.

Question 1: If we look at this question and use common sense, we know that an image can be both abstract and visual. The word "versatile" in the question also helps us find the answer.

Question 2: Ask yourself "Why might a child remember a signal longer than a dog?" B, C, and D don't seem like reasonable answers. For answer B, what does it mean for a human's brain to evolve more quickly? This answer is somewhat unintelligible. Answer C compares the age of a human child with a dog in terms of memory as if they were equivalent. This doesn't seem to be reasonable. At best, it calls for outside information about a dog's ability to remember based upon its age. For answer choice D, the question doesn't say anything about vision. Where does color-blind come in. This is a beyond.

Question 3: Notice that the question asks what is meant by the quote. For this type of question, just match the answer to the quote. Answer B is a paraphrase of the quote. Sidney himself is superfluous information.

Question 4: This is the same type of question as the last. Match the answer to the quote. Notice answer choice D. This is for those who want to see things in black and white, and take the quote very literally. It also does not match the quote.

Question 5: What would weaken the claim that the use of symbolic imagery is unique to humans? An example of a non-human using symbolic imagery. A is correct.

Question 6: Here we are asked to interpret a paraphrase. Just match the paraphrase to the answer choice. "bringing reality into existence" is the same as "creating the future".

Question 7: This is difficult to answer without the passage. However, look at the other questions. Ask yourself, "What is the main idea of this passage?" It is certainly about images, symbols, and language. Which answer fits most closely? Notice that the word image is in all the answers except the correct one. This makes choices B, C, and D simpletons. A is correct.

Question 8: The word "urgently" helps to narrow down the choices to B and D. It is difficult to choose between these two without reading the passage.

Question 9: The question appears to be impossible to answer without reading the passage; however, we will find that it is easy to answer after we answer the other questions. We'll come back to it.

Question 10: This is also a difficult question to answer before we answer the others. We'll come back to it.

Question 11: The word "Texas" should stand out here. What is special about Texas? There must be something special about location and Folsom hunter sites. Looking at the answer choices, "Colorado" also stands out. What is special about Colorado? D seems like a pretty good answer. There is certainly no reason to choose any other answer. But then again, there doesn't seem to be any reason not to either. We'll come back to this one too.

Question 12: Here it is again; location! This question gives us another clue. Apparently the author believes that tools are made from material found near a site because if tools were found that were not made from material near the site, this would somehow weaken the author's argument. Here, it makes sense that if the author argued for either II or III, both would be weakened. There doesn't seem to be any reason why choice I would be weakened. B or D must be the answer. But wait, if the author thought that Folsom hunters did travel great distances, then they might have carried tools with them, and III would not be weakened. Therefore, D must be the answer.

Question 9 revisited: Now we know that the answer to question 9 must be D.

Question 10 revisited: Now we know that D must be wrong. Notice the word "single" emphasizing the importance that the tools appear together. Choice B also addresses this togetherness with the word "communal". A and C don't seem to have a mechanism which would explain them. The correct answer is B.

Question 11 revisited: Clearly, if tools must be made from nearby materials, "tools made from Colorado flint" would not be found at a site in "northern Texas". D is correct.

Question 13: This question even tells you part of the answer to question 12. Answer A just confirms what we've already discovered. A is correct.

Question 14: This question can be answered using the other questions as background. The passage is about the Folsom hunters. Only D incorporates this into its answer. The main idea is always the best choice.

Question 15: The question asks for something that would prove (show) that the hunters killed the bison that they ate. In other words, the question wants something that would show that they didn't scavenge the bison after finding them dead, but they actually killed them. Choice A shows that the bison were killed with tools made by the hunters. Choice A is also the only choice that seems consistent with a main idea that apparently has to do with tools.

Question 16: Notice that the question emphasizes small versus large. The only answer choice that addresses this emphasis is choice C. If the game were small, the hunters could have carried their prey to the site.

Question 17: If you know that Galileo was forced by the church to recant his theories, you know that A is wrong.

This information also seems to suggest C or D. We'll come back to this one.

Question 18: This is just common sense. Now we know what the passage is about.

Question 19: Certainly B is wrong. The problem with both A and D is that the example includes a 'good thing' about science, the abandonment of eugenics, and a bad thing about science, Hitler's use of it. (Eugenics is the creation of the perfect race through breeding.) A and D suggest that both are bad, so they are the wrong answer choices. C is correct.

Question 20: This is a perfect example of a non-MCAT verbal thought: "science is an inexorable march toward truth." Verbal is ambiguous, not absolute. This is extremely unlikely to be the belief of an author in the verbal section. A is clearly the correct answer because it is so MCAT-like.

Question 21: The main idea is about the relationship between social events and science. Clearly B is correct. B is also very MCAT-like in its non-absolutism.

Question 22: B is the main idea; it must be correct. A and D are beyonds. C is not MCAT-like.

Question 23: C and D don't make sense; they seem to contradict the logic of the question. A answers the question better than B because B has science affecting society as opposed to the way the question has society affecting science.

Question 17 revisited: We were stuck between C and D. Unfortunately, we're still stuck. We'll have to just guess, or read the passage. We still get at least an MCAT score of 12.

The correct answers are: 1. B, 2. A, 3. B, 4. B, 5. A, 6. D, 7. A, 8. D, 9. D, 10. B, 11. D, 12. D, 13. A, 14. D, 15. A, 16. C, 17. C, 18. C, 19. C, 20. A, 21. B, 22. B, 23 A.

This exercise is not to convince you not to read the passage. You should always read the passage. It should show you that there is a large amount of information in the questions and answer choices. If you scored higher without reading the passage, then you probably haven't been taking advantage of the wealth of information in the questions and answer choices.

2-5 Marking Your Test to Improve Your Score

As you review the possible answer choices to a question, you should mark the letters of each answer choice with one of four symbols meaning: 1. absolutely incorrect; 2. probably incorrect; no idea; and 3. possibly correct. A diagonal line through a letter indicates that you have dismissed that answer as absolutely incorrect. A small x to the left of the letter indicates that you have a feeling that the answer is wrong, but you are not certain. A small horizontal line to the left of the letter indicates that you have no inclination as to whether the answer choice is right or wrong. A small circle to the left of the answer choice indicates that you like the answer but are not certain that it is correct. Below is an example:

A. Absolutely incorrect
× **B.** Probably incorrect
– **C.** No idea
∘ **D.** Possibly correct

This marking system saves time by helping you keep track of your thoughts concerning each answer choice. Recording your impressions next to each letter also helps guide your instincts providing for more accurate guesses.

2-6 When to Bubble

When you decide upon an answer, circle the correct letter and bubble in the answer on your test immediately. DO NOT wait till you have answered several questions and then bubble in several answer choices at a time. This is the most common way that bubbling errors occur. Bubbling in answer choices is NOT a productive way to spend your 5 second break. Bubble as you go, one answer at a time. This is the fastest, most accurate and efficient method.

Lecture 3

The Main Idea

3-1 The Main Idea

The main idea is a summary of the passage in one or two sentences. It should reflect the author's opinion (if presented or implied), and it should emphasize minor topics to the same extent as they are emphasized in the passage. It is not a list of topics discussed in the passage nor an outline of those topics. It is a statement about the passage topics, and includes the author's opinion.

In one form or another, 90% of the Verbal Section questions will concern the main idea. Notice that the main idea cannot be found by going back to the passage and searching for details. You must concentrate on the main idea while you read the entire passage. If you read for detail, if you try to remember what you have read rather than process what you are reading, you will have to guess at 90% of the questions.

It is important to have a clear concept of the main idea before reading any questions. MCAT Verbal section questions are designed to take your inchoate thoughts concerning the passage and subtly redirect them away from the true main idea. Each successive question embellishes on insidious pseudo-themes steering unwary followers into an abyss from which there is no return. Like a faithful paladin, your clearly stated main idea unmasks these impostors and leads you toward the holy grail of Verbal Section perfection.

Writing the main idea on paper is an important step toward improving your ability to find the main idea; however, it requires too much time while taking the exam. Instead, the a few days after taking a practice exam, go back to each passage and write out the main idea. While taking the exam, make a 20 second pause after reading a passage, and construct the main idea in your head.

Most students resist writing out their main idea until they are halfway through the course and the materials. At this point they begin to realize how important the main idea is. Unfortunately, they must start from scratch and begin writing out the main idea with only four weeks until the MCAT. Don't do this. Start now by going back to used passages and writing out the main idea. It's very painful at first, but it will get easier, and it will dramatically improve your score.

3-2 Constructing the Main Idea

A good main idea can be formed as follows: 1. After reading the passage, write down the main topics. Each topic should be from one to four words. 2. From these topics, choose the most important ones two or three at a time, and write a short phrase relating them to each other and the passage. 3. Now connect the phrases into one or two sentences which still concern the most important topics but incorporate the other topics as well. Be sure to include the author's opinion if it was given or implied. Try to emphasize each topic to the same extent to which it was emphasized in the passage. This is your main idea.

3-3 Confidence

Often on the MCAT, passages seem incomprehensible. Don't get bent! Remember, most questions are answered correctly by 60% or more of test-takers, and only two or three are answered incorrectly by less than 40%, so no group of questions will be that difficult. Have the confidence to keep reading. **Don't reread a line or paragraph over and over until you master it.** If a line or paragraph is incomprehensible to you, then it is probably incomprehensible to everyone else, and understanding it will not help your score. Instead, continue reading until you get to something that you do understand. Just get the general sense of what the author is trying to say. Chances are good that this will be enough to answer all the questions. Remember, after you read the passage you have four tools beyond your understanding of the passage to help you answer the questions.

3-4 Know Your Author

You must become familiar with the author. Who is he or she? Is the author young or old; rich or poor; male or female; conservative or liberal? Do you love or hate this author? Take a guess. Create a picture of the author in your mind. Use your prejudices to stereotype the author. Your harsh judgment of the author is everything to understanding what he is trying to say. The better you understand the author, the easier the questions will be. Read with emotion and judge harshly.

Now that you know the author intimately, when you get to a question, ask yourself "If this author were right here in front of me, how would he answer this question?" The way that the author would answer the question, is the correct answer.

3-5 Ignore the Details and See the Big Picture

There is no reason to remember the details of a passage. They can be found in seconds, and are rarely important to answering a question. Instead, focus on the big picture. Ask yourself "What is the author trying to say to me? What's his beef?" The author's 'beef' will be the main idea, and the key to answering 90% of the questions.

STOP!

DO NOT LOOK AT THE FOLLOWING PASSAGE AND QUESTIONS UNTIL CLASS. IF YOU WILL NOT BE ATTENDING CLASS, READ THE PASSAGE IN THREE MINUTES AND ANSWER THE QUESTIONS WHICH FOLLOW.

Passage

It is roughly a century since European art began to experience its first significant defections from the standards of painting and sculpture that we inherit from the early Renaissance. Looking back now across a long succession of innovative movements and stylistic revolutions, most of us have little trouble recognizing that such aesthetic orthodoxies of the past as the representative convention, exact anatomy and optical perspective, the casement-window canvas, along with the repertory of materials and subject matters we associate with the Old Masters—that all this makes up not "art" itself in any absolute sense, but something like a school of art, one great tradition among many. We acknowledge the excellence which a Raphael or Rembrandt could achieve within the canons of that school; but we have grown accustomed to the idea that there are other aesthetic visions of equal validity. Indeed, innovation in the arts has become a convention in its own right with us, a "tradition of the new," to such a degree that there are critics to whom it seems to be intolerable that any two painters should paint alike. We demand radical originality, and often confuse it with quality.

Yet what a jolt it was to our great-grandparents to see the certainties of the academic tradition melt away before their eyes. How distressing, especially for the academicians, who were the guardians of a classic heritage embodying time-honored techniques and standards whose perfection had been the labor of genius. Suddenly they found art as they understood it being rejected by upstarts who were unwilling to let a single premise of the inherited wisdom stand unchallenged, or so it seemed. Now, with a little hindsight, it is not difficult to discern continuities where our predecessors saw only ruthless disjunctions. To see, as well, that the artistic revolutionaries of the past were, at their best, only opening our minds to a more global conception of art which demanded a deeper experience of light, color, and form. Through their work, too, the art of our time has done much to salvage the values of the primitive and childlike, the dream, the immediate emotional response, the life of fantasy, and the transcendent symbol.

In our own day, much the same sort of turning point has been reached in the history of science. It is as if the aesthetic ground pioneered by the artists now unfolds before us as a new ontological awareness. We are at a moment when the reality to which scientists address themselves comes more and more to be recognized as but one segment of a far broader spectrum. Science, for so long regarded as our single valid picture of the world, now emerges as, also, a school: a *school of conscious-ness,* beside which alternative realities take their place.

There are, so far, only fragile and scattered beginnines of this perception. They are still the subterranean history of our time. How far they will carry toward liberating us from the orthodox world view of the technocratic establishment is still doubtful. These days, many gestures of rebellion are subtly denatured, adjusted, and converted into oaths of allegiance. In our society at large, little beyond submerged unease challenges the lingering authority of science and technique, that dull ache at the bottom of the soul we refer to when we speak (usually too glibly) of an "age of anxiety," an "age of longing."

GO ON TO THE NEXT PAGE.

Answer the following questions without going back to the passage. If you don't know the answer, guess.

YOU MAY NOT LOOK AT THE PASSAGE!

- Is the author male or female?
- Does the author have long or short hair?
- How old is the author?
- What political party is the author a member of?
- Would the author prefer a wild party or a night at the opera?
- Do you think you would like the author?
- What does the author do for a living?

These are the types of questions that you should be able to answer with prejudice if you have read the passage the way you should. If you can answer these questions, you have compared the author to people of your past and categorized the author accordingly. This means that you have a better understanding of who the author is, and how he would answer the MCAT questions about his own passage.

The previous questions were asked to make you realize how you should be trying to understand the author. You should not be asking yourself these questions on a real MCAT. Here are some questions that you should ask yourself on a real MCAT:

YOU MAY NOT LOOK AT THE PASSAGE!

- If the author were sitting in front of you, would he or she want to discuss science or art?
- What emotion, if any, is the author feeling?
- Is the author a scientist?
- Is the author conservative, liberal, or somewhere in the middle?

The answers to these questions are unequivocal. This author is discussing science, not art. Art is used as a lengthy, nearly incomprehensible introduction to make a point about science. The author doesn't even begin discussing the main idea until the beginning of the third paragraph. "In our own day, much the same sort of turning point has been reached in the history of science." When you read this, you should have been startled. You should have been thinking "Where did science come from? I thought we were talking about some esoteric art history crap that I really wasn't understanding." This one sentence should have said to you "Ahaa! That other stuff was appetizer, now the author is going to discuss his real interest." Notice that it is at the beginning of the third paragraph that the writing actually becomes intelligible. In other words, the second two paragraphs are much easier to read. This is because the author is interested in this topic and knows what he wants to say. The art stuff was a poorly written introduction and the author had not thought it through with any clarity. If you spent lots of time rereading the first two paragraphs, trying to master them, you wasted your time. The author didn't even master them; how could you?

The author is frustrated and possibly even bitter. He is so angry, that he is name-calling. For instance, he calls the scientific community "the technocratic establishment". The tone of the passage is like that of a whining child. He blames scientists for being too conservative and thus creating "an age of anxiety", as if the anxiety of most people would be relieved if scientists were less practical. In the last sentence, he even blames us, his reader, for not taking *his* issue more seriously. The author is positively paranoid. Notice that his adversaries move against him "subtly" as if trying to hide their evil intentions. They take "oaths of allegiance" like some NAZI cult. This is way overdone when you consider that the guy's only complaint is that science isn't liberal enough in its approach.

The author is certainly not a scientist. First of all, he writes like a poet not a scientist: "orthodox world view of the technocratic establishment", "subterranean history of our time", "gestures of rebellion subtly denatured". Secondly, his whole point is that he is upset with scientists. (An entire separate argument can be made that his point results from his misunderstanding of how science progresses.) And finally, he talks like a member of some pyramid cult, not a scientist: "alternative realities" and "ontological awareness". This author probably flunked high school physics and just can't get over it.

The author is certainly liberal, or anti-establishment. He talks about "liberating us" and "rebellion" among other things.

Now, with this understanding of the author, answer the questions on the next page.

YOU MAY NOT LOOK AT THE PASSAGE!

1. The author believes that in "the subterranean history of our time" (line 51-52) we find the beginnings of a:

A. renewal of allegiance to traditional values.
B. redefinition of art.
C. redefinition of science.
D. single valid picture of the world.

2. The author compares art and science mainly in support of the idea that:

A. the conventions of science, like those of art, are now beginning to be recognized as but one segment of a far broader spectrum.
B. aesthetic orthodoxies of the past, unlike scientific orthodoxies of the present, make up only one tradition among many.
C. artistic as well as scientific revolutionaries open our minds to a more global conception of art.
D. artists of the past have provided inspiration to the scientists of the present.

3. The two kinds of art discussed in the passage are the:

A. aesthetic and the innovative.
B. dull and the shocking.
C. traditional and the innovative.
D. representative and the traditional.

4. The author's statement "How far [new perceptions of science] will carry toward liberating us from the orthodox world view of the technocratic establishment is still doubtful" (lines 52-54) assumes that the:

A. technocratic establishment is opposed to scientific inquiry.
B. traditional perception of science is identical to the world view of the technocratic establishment.
C. current perceptions of science are identical to those of art.
D. technocratic establishment has the same world view as the artistic revolutionaries of the past.

5. Which of the following concepts does the author illustrate with specific examples?

A. Scientific innovations of the present
B. Scientific innovations of the past
C. Aesthetic innovations of the present
D. Aesthetic orthodoxies of the past

6. The claim that the unease mentioned in line 57 is "submerged" most directly illustrates the idea that:

A. our great-grandparents were jolted by the collapse of academic certainty.
B. we have grown accustomed to the notion that there is more than one valid aesthetic vision.
C. so far, new perceptions of science are only fragile and scattered.
D. the authority of science is rapidly being eroded.

7. Based on the information in the passage, the author would most likely claim that someone who did NOT agree with his view of science was:

A. dishonest.
B. conformist.
C. rebellious.
D. imaginative.

8. Based on information in the passage, which of the following opinions could most reasonably be ascribed to the author?

A. It is misguided to rebel against scientific authority.
B. The world views of other disciplines may have something valuable to teach the scientific community.
C. Art that rebels against established traditions cannot be taken seriously.
D. The main cause of modern anxiety and longing is our rash embrace of new scientific and artistic theories.

9. Adopting the author's views as presented in the passage would most likely mean acknowledging that:

A. it is not a good idea to accept traditional beliefs simply because they are traditional.
B. we must return to established artistic and scientific values.
C. the future is bleak for today's artists and scientists.
D. the scientific community has given us little of benefit.

STOP. IF YOU FINISH BEFORE TIME IS CALLED, CHECK YOUR WORK. YOU MAY GO BACK TO ANY QUESTION IN THIS TEST BOOKLET.

Don't worry about the correct answers yet.

YOU MAY NOT LOOK AT THE PASSAGE YET!

The first thing to notice is that only question 5 requires any information from the first two paragraphs, and question 5 was a question about detail, not concept. This is because the first two paragraphs are not about the main idea.

The second thing to notice is that none of the questions require us to go back to the passage, even though some refer us to specific line numbers. All but question 5 are answerable directly from the main idea. Question 5 is a detailed question, but before you run back to the passage to find the answer, look at the possibilities. The chances are that you remembered Raphael and Rembrandt from the first paragraph. These are specific examples of "aesthetic orthodoxies of the past".

Notice that many of the questions can be rephrased to say "The author thinks ________." This is typical of an MCAT passage, and that's why you must "know your author".

Question 1: Forget about the quote for a moment. Simplify the question to say "The author thinks that we find the beginnings of a:" Answer C is the main idea. Certainly the author would disagree with A, B, and D.

Question 2: "The author thinks:" that science is like art, and that conventions of both are but part of a larger spectrum. B says science is not like art; the opposite of what the author thinks. C says that scientific revolutionaries are changing science; the author is frustrated because this is not really happening. D says scientists of the present are opening their minds to new ideas; the author complains that they are not.

Question 3: The main idea of the passage contains the theme of traditional vs. innovative.

Question 4: Ignore the quotes until you need them. Without the quotes, the questions says "The author's statement assumes that the:" In other words, "The author thinks ________." C and D are exactly opposite to what the author thinks. Answer A plays a common game on the MCAT. They take the author's view too far. They want you to think "the author doesn't like the scientists; therefore, he thinks the scientists can't even do science." Even this author wouldn't go that far. A is incorrect. Answer B requires you to realize that the "technocratic establishment" is conservative.

Question 6: Answer D is out because it disagrees with the main idea, and C is the only answer that supports the main idea. However, this question is best answered by comparing the answer choices with the question. The question asks for an example of "submerged unease". "jolted" in answer choice A certainly doesn't describe submerged unease. "grown accustomed" in answer choice B certainly does not describe submerged unease. Answer choice C could describe submerged unease, and it does describe the main idea. It is the best answer.

Question 7: The author is rebellious and imaginative. If you disagree with him, he thinks you are a conformist, which, by the way, is worse than dishonest as far as he's concerned.

Question 8: "The author thinks ________." The whole point of the intro is to say that the scientific community should learn from the discipline of art.

Question 9: "The author thinks ________." The author is a rebel. He thinks you should always question authority. Notice choice D is another example of taking things too far. No sane individual could argue that science has provided little benefit. Answer choice C would be incorrect even if it had not included 'artists'. It would have been too extreme.

NOW YOU MAY LOOK AT THE PASSAGE.

Hopefully, we have demonstrated the power of knowing the author and understanding the main idea. Remember to use all four of your tools, and, most importantly, read and answer questions with confidence.

If you have problems, go back to the basics of this manual. Figure out what part of our strategy and tactics you aren't using, and use it.

How to Study for the Verbal Reasoning Section

4-1 How to Study for the Verbal Reasoning Section

One might think that studying the correct answers to many verbal questions in order to discover why they are correct would be a helpful exercise toward improving your score. On the contrary, it's probably a waste of your precious time. It is rarely useful to go back to old tests and learn the logic used to explain why the correct answers are correct. By doing so, you may learn something about the topic of the passage, but you do not learn what you can do differently next time to improve your score. Since most explanations justify answers by pointing to a specific place in the passage that is claimed to support or even prove the correct answer, such practices can even lower your MCAT score by giving you the false impression that answers can be found in a specific place in the passage. Most MCAT answers require an understanding of the passage as a whole and cannot be proven correct by reading from one place in the passage. In most verbal materials, explanations tend to be too brief and not particularly insightful. It is doubtful that reading them will increase your reading comprehension skills.

One method for increasing your reading comprehension skills is to join a book club or organize a reading group, and discuss things that you have read. This is not particularly practical for most premeds, and even this idea is only effective if there are strong, insightful readers in your club or group. The often posited advice of reading lots of magazine and newspaper articles on your own is a significantly less effective method for improving your reading comprehension skills. At the very least, you should be spending your reading time doing verbal passages followed by questions and not just articles without questions.

The most effective method of study to improve your MCAT verbal score is to do the following:

1. Take a verbal test under strict timed conditions and score yourself.
2. Take a break from verbal for at least one day.
3. Take the set of questions for the first passage in the verbal exam that you recently finished and examine the questions and each answer choice as if you had never read the passage, as was done in Lecture 2 of this book. If this step takes you less than 30 minutes per passage, then do it again because you missed quite a bit.
4. Repeat step 3 for each passage.
5. Take a break from verbal for at least one day.

6. Carefully read the first passage in the same verbal test, and write out a precisely worded main idea in one or two complete sentences being certain that your main idea expresses the author's opinion or stance on the issues.
7. Match your main idea to each question and all the answer choices and see what insights you gain into answering MCAT questions as was done in Lecture 3 of this book.
8. Repeat steps 6 and 7 for each passage.

STOP!

DO NOT LOOK AT THESE EXAMS UNTIL CLASS.

30-MINUTE IN-CLASS EXAM FOR LECTURE 1

Passage I (Questions 1–7)

Philosophers Immanuel Kant and David Hume both spent their professional careers searching for a universal principle of morality. Considering that they began their searches with seemingly irreconcilable ideas of where to look, the similarity in the moral systems they constructed is surprising. ...

Hume decided at the outset that a moral system must be practical, and maintained that, since reason is only useful for disinterested comparison, and since only sentiment (emotion) is capable of stirring people to action, the practical study of morality should be concerned with sentiment. Hume begins with the assumption that whether something is judged moral or praiseworthy depends on the circumstances. He says, "What each man feels within himself is the standard of sentiment." ...

By contrast, Kant begins by assuming that, while some time should be devoted to studying practical morality ("ethics"), it is also valuable to have an *absolute* system of morality based solely on reason, to be called "metaphysics". That forms the core of his laborious exploration of pure logic, called *Grounding for the Metaphysics of Morals*.

For both authors, the problem of subjectivity threatened to prevent unbiased analysis of morality. So both invented systems of "moral feedback," in which the philosophic actor tries to imagine the results of his actions as if someone else were performing them. Still, the final evaluation of the action's worth must, in the end, be subjective. Kant's ultimate standard of morality is the "categorical imperative". It is phrased, "[To follow your] duty, act as if the physical act of your action were to become a universal law of nature." This means that, before one does anything, one should forget one's own motives for a moment, and ask if he would want everyone to do as he does. If the answer is no, then his subjective desire is different from his objective assessment, and the action is contrary to duty.

Hume says that moral actions are those that create agreeable sentiments in others, as well as in oneself. You can therefore judge what kind of sentiments your actions may cause others to feel by imagining someone else performing that action, and thinking about what kind of sentiments it would inspire in you. To make both of these constructions possible, there is a notion of some kind of uniformity in people's reason or emotion in both works; Kant's reason for using pure logic was precisely to bypass empirical differences between people and individual circumstances, and ... he has the belief that people using only logic must inevitably reach the same conclusions about the morality of their own actions. Hume also admits that "the notion of morals implies some sentiment common to all mankind." ...

Some Marxist theorists have speculated that Marx would tar both Kant and Hume as "bourgeois" philosophers. Answering why is an extremely difficult question. ... It may lie in this sentence from Marx's *Manuscripts of 1844*: "The interests of the capitalist and those of the workers are therefore, *one and the same*, assert the bourgeois economists." ... In the directly preceding passage, Marx explained how capitalists try to mask the class struggle and exploitation inherent in capitalist production, by monopolizing cultural institutions to establish hegemonic control over the working class. This is why, he explains, workers are actually conditioned to be grateful to the factory owner for allowing them to produce goods, only to have these taken away and sold. Based on these writings, Marx would probably see any system that sought out a universal theory of morality as ignoring the opposing economic classes in society, and easily adapted to give the workers a false perception of the unitary interest of them and their oppressors.

1. The word *tar* (line 53) is used in the sense of:

- **A.** asphalt.
- **B.** suggest.
- **C.** label.
- **D.** stick.

2. According to the passage, Marx would have disagreed with Kant and Hume over which of the following ideas?

- **A.** What each man feels within himself is the standard of sentiment.
- **B.** There is uniformity in people's reason or emotion.
- **C.** The interests of the capitalists and the workers are one and the same.
- **D.** Morality is dependent upon the class struggle.

GO ON TO THE NEXT PAGE.

3. Assume that a universal principle of morality can be proven to exist. Which of the following hypotheses does this assumption suggest?

A. The author is correct; despite their genesis, it is not surprising that Kant and Hume constructed similar systems.
B. The author is correct; Marx, Hume, and Kant all constructed similar systems.
C. The author is incorrect; Marx, Hume, and Kant did not all construct similar systems.
D. The author is incorrect; despite their genesis, it is not surprising that Kant and Hume constructed similar systems.

4. According to the author, in creating his moral system, Hume equated:

A. circumstances with disinterested comparison.
B. circumstances with absolutism.
C. practicality with stirring people to action.
D. practicality with reason.

5. Based upon passage information, Marx most likely believed that:

A. there is no universal theory of morality.
B. philosophers were part of the bourgeois.
C. the workers could be easily fooled.
D. Kant's book supported capitalist exploitation.

6. Based upon passage information, Kant's system of "moral feedback" (line 25) differed from Hume's in that it might result in a situation wherein:

A. one realized that his action might be 'right' so long as it didn't become a universal law of nature.
B. one realized that a universal law of nature was unnecessary in determining duty.
C. one realized that his action might be 'wrong' even though it created agreeable sentiments in others, as well as in oneself.
D. one realized that his action might be illogical if sentiment was not further considered.

7. According to the passage, Hume's ideas evolved to the point where he:

A. realized that reason was an inseparable part of a universal system of morality.
B. was considering the sentiments of others as well as himself.
C. chose to essentially agree with Kant on a universal system of morality.
D. decided that sentiment without action was a necessary component of his morality system.

GO ON TO THE NEXT PAGE.

Passage II (Questions 8-14)

In the Bible, when Jesus Christ was queried by skeptical Israelites on how Christians could continue to pay taxes to support the (pagan) Roman governors of Israel, he counseled, "Render unto Caesar what is Caesar's; render unto God what is God's". The basic meaning of this guidance was that Christians may practice their faith while co-existing with the secular government. ... Later Christian theorists followed this example, often urging peaceful co-existence even with governments which violated every precept of Christian teachings. ...

The early Catholic bishop (St.) Augustine of Hippo and the Protestant dissident leader Martin Luther both advocated submission to the rule of tyrants, employing the analogy of a *dual* city or government, where earthly rule is often oppressive, yet is balanced by the assurance of brotherly love in the kingdom of heaven. ...

According to Luther, true Christians should be willing to suffer persecution, without seeking to resist it by the anti-Christian methods of taking up arms in violent revolt, or seeking redress in the courts of the unbelievers. Luther even took the analogy so far as to suggest that it is the Christians themselves who most benefit from harsh secular laws, which protect *them* from exploitation and persecution by false Christians and heathen. In fact, when commenting on popular "Christian" uprisings, Luther lays more blame on self-righteous Christian rebels (such as various German peasant rebels of the age) than upon their oppressors, teaching that, regardless of their rulers' faults, rebels immediately cease to be Christians upon taking up arms, and incur further displeasure from God by blasphemously arrogating His name and scriptures for an un-Christian cause. Luther claims that this is furthermore exacerbated because what the rebels often want is material benefit rather than religious freedom. ... According to Luther, it is the role of God alone to punish rulers, and rebels' usurpation of that authority for themselves adds another sin to the list of charges against them; they effectively revolt against God's justice.

St. Augustine would have agreed. His *Confessions* and other theological writings maintained that God grants earthly rule to Christians and pagans alike, but that His inscrutable will is always just. In the case of a revolt, Luther claimed, *both* sides inevitably incur divine punishment, since "God hates both tyrants and rebels; therefore He sets them on each other". Luther's vision of God raising up the peasants to punish their tyrants perfectly matches Augustine's frequent portrayal of the Germanic "barbarians" who finally sacked Rome as the brutal instruments of God, sent to crush the corrupt Romans' arrogance. Augustine never considers the individual fate of these living tools, but Luther maintains that God sends the devil to stir them up with lies, and afterwards they go to eternal torment.

Both of these theologians would hold that uprisings generally occur for the wrong reasons (i.e. worldly ambitions), because no tyrant can keep a true Christian from salvation, which should be all Christians' only concern in life. For Luther, salvation lies in faith emanating from personal understanding of Biblical teachings, and "it is impossible that anyone should have the gospel kept from him...for it is a public teaching that moves freely." In Augustine's understanding, salvation is granted through the mercy of God, who sends hardship and death to test men's faith. The true Augustinian Christian would maintain his faith through any ordeal, and even if his body perishes, God will save him for his conviction.

8. Which of the following statements, if true, would most directly *challenge* the principles of Martin Luther?

- **A.** The Bible's Old Testament refers to a period before the birth of Jesus.
- **B.** The Bible alone contains only a small part of what Jesus intended for his followers.
- **C.** The German "barbarians" who sacked Rome had been previously converted.
- **D.** Augustine's understanding of salvation granted through the mercy of Christ was flawed.

9. Some theologians believe that killing and violence are acceptable when used in self-defense. An appropriate clarification of the passage would be the stipulation that:

- **A.** both Luther and Augustine would have disagreed with this belief.
- **B.** both Luther and Augustine would have agreed with this belief.
- **C.** only Augustine might have agreed with this belief.
- **D.** only Luther might have agreed with this belief.

10. If the information in lines 38-55 is correct, one could most reasonably conclude that, compared to Luther, Augustine was:

- **A.** much more reasonably inclined.
- **B.** more prepared to define God's will.
- **C.** less eager to send people to eternal torment.
- **D.** less willing to announce God's final judgment on those who had sinned.

GO ON TO THE NEXT PAGE.

11. The author's attitude toward the theories of Augustine and Luther in the passage is most accurately described as:

A. disapproving.
B. mistrustful.
C. neutral.
D. favorable.

12. What is the meaning of the phrase; "The true Augustinian Christian would maintain his faith through any ordeal, and even if his body perishes, God will save him for his conviction" (lines 61-64)?

A. This Christian would end up in heaven because of his beliefs.
B. God would save this Christian for judgment at the end of the Christian's ordeal.
C. God would save this Christian from his ordeal and judge him.
D. It was not necessary for this Christian to die for him to be convicted.

13. The author's primary purpose in the passage is apparently:

A. to clarify the differences between the ways in which the early Catholics and Protestants dealt with persecution.
B. to justify the persecution of early Christians by secular governments.
C. to consider the similarities between the ways in which the early Catholics and Protestants dealt with persecution.
D. to question the passive practices of the early Catholics and Protestants when faced with persecution.

14. What is the most serious apparent weakness of the information described?

A. While implying that Christians may coexist with a secular government, it differentiates between Catholics and Protestants.
B. While implying representation of Augustine and Luther, its conclusions are based primarily on information according to Luther.
C. While implying representation of all Christian theorists, only Augustine and Luther are mentioned.
D. While implying agreement between Augustine and Luther, their attitudes were clearly opposed.

GO ON TO THE NEXT PAGE.

Passage III (Questions 15-21)

Of all the bizarre and melancholy fates that could befall an otherwise ordinary person, Mary Mallon's has to be among the most sad and peculiar. …Like millions before and since, she came to this country from Ireland, seeking a better life. Never "tried" in any sense, instead, she was forced by public health officials to live for a total of 26 years on a tiny island in the East River, isolated from and shunned by her fellow humans. And, while she was not the only one of her kind, her name became synonymous with disease and death. She was Typhoid Mary, and her story really begins on Long Island.

In the summer of 1906, Mallon, was working as a cook for a wealthy New York banker, Charles Henry Warren, and his family. The Warrens had rented a spacious house in Oyster Bay, "in a desirable part of the village," for the summer. From August 27 to September 3, six of the 11 people in the house came down with typhoid fever, including Mrs. Warren, two daughters, two maids and a gardener. Two investigators were unable to find contaminated water or food to explain the outbreak. Worried they wouldn't be able to rent the house unless they figured out the source of the disease, the owners, in the winter of 1906, hired George Soper, a sanitary engineer.

Soper soon dismissed "soft clams" and other potential contaminants as the cause and began to focus on the family. He later wrote, "It was found that the family had changed cooks about three weeks before the typhoid epidemic broke out ... She remained with the family only a short time, leaving about three weeks after the outbreak occurred ... [and] seemed to be in perfect health." Soper became convinced that this woman was a healthy carrier of the disease, and, in so doing, was the first to identify a healthy typhoid carrier in the United States. Although his deduction was undoubtedly brilliant, his handling of Mallon was not.

Soper tracked Mallon down to a home on Park Avenue in Manhattan where she was a cook. Appearing without warning, Soper told her she was spreading death and disease through her cooking and that he wanted samples of her feces, urine, and blood for tests. In a later description, Soper wrote, "It did not take Mary long to react to this suggestion. She seized a carving fork and advanced in my direction. I passed rapidly down the narrow hall through the tall iron gate."

Convinced by Soper's information, the New York City health inspector in March 1907, carried Mallon off, screaming and kicking, to a hospital, where her feces did indeed show high concentrations of typhoid bacilli. She was moved to an isolation cottage on the grounds of the Riverside Hospital, between the Bronx and Rikers Island.

... She stayed there for three years, in relative isolation. It was during that time that she was dubbed Typhoid Mary. Mallon despised the moniker and protested all her life that she was healthy and could not be a disease carrier. As she told a newspaper, "I have never had typhoid in my life and have always been healthy. Why should I be banished like a leper and compelled to live in solitary confinement ...?"

After a short period of freedom in which Mallon failed to comply with the health inspector's requirements, she was eventually sent back to North Brother Island, where she lived the rest of her life, alone in a one-room cottage. In 1938 when she died, a newspaper noted there were 237 other typhoid carriers living under city health department observation. But she was the only one kept isolated for years, a result as much of prejudice toward the Irish and noncompliant women as of a public health threat.

15. The author probably mentions that Mallon was "never 'tried' in any sense" (line 5) in order:

- **A.** to demonstrate the power of the wealthy at that time.
- **B.** to provide a comparison with people who have actually committed a crime.
- **C.** to illustrate the persistence of Soper's investigations.
- **D.** to support the claim that she deserved at least a hearing.

16. According to the passage, the first two investigators were unable to find the cause of the outbreak (lines 19-20). The information presented on typhoid makes which of the following explanations most plausible?

- **A.** They focused too closely on the "soft clams" that Soper later discredited.
- **B.** Typhoid is not really passed through contaminated food or water.
- **C.** They never considered that typhoid could be carried by a healthy person.
- **D.** By this time, Mallon was no longer employed by the Warren family.

17. The author's argument that Mallon's isolation was "a result as much of prejudice ... as of a public health threat" (lines 64-65) is most weakened by which idea in the passage?

- **A.** Mallon's primary occupation was as a cook.
- **B.** Mallon did not believe that she was a carrier of the disease.
- **C.** Mallon would not abide by the health inspector's requirements.
- **D.** Mallon actually was the source of the typhoid outbreak in the Warren home.

18. Which of the following statements is the most reasonable conclusion that can be drawn from the author's description of the typhoid outbreak in the house at Oyster Bay?

- **A.** The Warren family did not hire Soper.
- **B.** The two investigators were hired by the Warrens.
- **C.** The Warren family hired Soper.
- **D.** The owners were anxious to sell the house.

19. Passage information indicates that which of the following statements must be true?

- **A.** Mallon had probably not infected anyone prior to the Warren family.
- **B.** Mallon was almost certainly not washing her hands prior to preparing the Warren's meals.
- **C.** Being labeled 'Typhoid Mary' by the press was the primary reason for her confinement.
- **D.** The health inspector was doubtless prejudiced toward the Irish.

20. According to passage information, Mallon worked for the Warren family for approximately:

- **A.** two weeks.
- **B.** three weeks.
- **C.** five weeks.
- **D.** six weeks.

21. The contention that in "1938 when [Mallon] died, ... there were 237 other typhoid carriers living under city health department observation. But she was the only one kept isolated for years" (lines 61-65), can most justifiably be interpreted as support for the idea that:

- **A.** Mallon was unfairly treated by the city health department.
- **B.** Mallon's isolation might have stemmed from the health department's early ignorance of the disease.
- **C.** The "other" 237 typhoid carriers were all kept isolated at one time or another.
- **D.** The "other" 237 typhoid carriers were much like Mallon.

STOP. IF YOU FINISH BEFORE TIME IS CALLED, CHECK YOUR WORK. YOU MAY GO BACK TO ANY QUESTION IN THIS TEST BOOKLET.

30-MINUTE IN-CLASS EXAM FOR LECTURE 2

Passage I (Questions 22–28)

Polling research shows that the ideal speaking voice should be clear and intelligible, of moderate volume and pace, and inflected to suggest the emotions expressed. To suggest credibility, the voice's tone should be pitched as low as is naturally possible... A low pitch is desirable in both genders, since it is popularly associated with truth-telling. However, an artificially lowered voice can sacrifice intelligibility, which is irritating to listeners. Thus, speakers should experiment to find their optimal level, which will be at their lowest intelligible pitch. ... To deepen the pitch, speakers should make an extra-deep inhalation before speaking, and then exhale fully as they speak.

Habits to be avoided (because they irritate most listeners) include monotony, mumbling, grating, pretension, high-pitched whining, and breathiness. ... Among the best practitioners of mainstream vocal "propriety" are famous news anchors, like Walter Cronkite, Dan Rather, and Jane Pauley, for whom vocal image is a key component of their job success. However, everyone who communicates should be aware that their voice is a critical component of their audience's perceptions of them, comprising about 38% of the overall impression imparted by their presentation. (By comparison, appearance accounts for about 50% of the speaker's impact, and the quality of content accounts for a mere 6%.) ...

For those not born with naturally pleasant voices, or worse still, those with naturally unpleasant ones, speech training can be invaluable in improving impressions. No voice teacher is necessarily required, since practice alone can produce significant improvements. However, some special equipment is needed. Because talking always causes cranial resonance, which distorts the speaker's hearing, no one can hear what his voice really sounds like to a listener. Thus, some sort of tape recorder or other feedback is a virtual necessity.

Speaking begins with breathing, since speech is just exhaled air that sets the vocal cords to resonating. If insufficient air is inhaled before speaking, the words formed must necessarily be strained and breathless. ... Diaphragmatic breathing results in a deeper voice than upper-lung breathing. People tend to stick out their chests and inhale shallowly with the upper lungs, resulting in a high-pitched voice, which must also be rapid to avoid running out of breath before the sentence ends. In diaphragmatic breathing, the lower abdomen moves out, inflating the bottom two-thirds of the lungs fully, but the shoulders do not rise. To practice switching from shallow breathing to deep, diaphragmatic inhalations, it helps to lie on the floor and breathe naturally, since diaphragmatic breathing is natural when prone. ... Increasing lung capacity will also deepen the voice and permit longer sentences without pausing. ... One method to increase lung capacity is to inhale fully, then count out loud slowly, while enunciating each number clearly, aiming for a count of 60.

Loudness, or volume, is distinct from pitch, though the remedy for overly soft-spoken people is similar. They can manage to speak louder by first inhaling more deeply, which allows them added lungpower to project their sentences. Alternately, they can pause more often, say fewer words in every breath, and thus leaving more air power for each. There are those who speak too softly not because of improper breathing, but due to psychological factors: they may be shy and not wish to be obtrusive, or may not hear that their words are too soft to be intelligible at a distance. ... For them, one useful exercise is to recognize the five basic volume levels (whisper, hushed, conversation, loud, and yelling) by practicing speaking a word in each of these modes. ...

22. According to the passage information, which of the following would be most likely, if a person, who was talking to you, attempted to make their voice sound unusually low?

- **A.** You might think that they were lying.
- **B.** They could be irritated with you.
- **C.** They might well sound monotonous.
- **D.** You could find them difficult to understand.

23. The author most likely believes that one of the main purposes of speaking, during a face-to-face meeting, should be to:

- **A.** convey a favorable impression.
- **B.** effectively transmit your ideas.
- **C.** gain leverage.
- **D.** communicate as naturally as possible.

GO ON TO THE NEXT PAGE.

24. The author provides a list of "habits to be avoided" (lines 13-15). Which of the habits would the suggestions in this passage not help a speaker to curb?

A. monotony
B. breathiness
C. pretension
D. high-pitched whining

25. The term ideal speaking voice (line 1) refers implicitly to a voice that is:

A. the most pleasant to listen to.
B. the most persuasive.
C. the least irritating.
D. the most natural.

26. Passage information indicates that a person speaking in a high-pitched voice might be doing all of the following EXCEPT:

I. Breathing with their upper lungs
II. Breathing deeply
III. Lying

A. I only
B. II only
C. III only
D. I and III only

27. Which of the following assertions is most clearly a thesis presented by the author?

A. Speakers can gain by improving his or her speaking voices.
B. The tone of the ideal speaking voice should be pitched as low as possible.
C. What you are saying is more important than how you are saying it.
D. Emotional inflections can be an irritating aspect of a speaker's voice.

28. The ideas discussed in this passage would likely be of most use to:

A. A doctor
B. A journalist
C. A radio show personality
D. A television commentator

GO ON TO THE NEXT PAGE.

Passage II (Questions 29–35)

A great deal of international conflict arises from border disputes. Throughout history, particularly along borders which have been "artificially" defined, rather than utilizing more natural pre-existing cultural and geographical demarcations, there has been a constant ebb and flow as nations have sought to consolidate their borders and their security. However, with ever-increasing economic disparity between many bordering countries, these conflicts have changed and now center more around issues of immigration. … Such situations are prevalent today in countries such as New Zealand, the Colombian-Peruvian border, and the U.S.'s Mexican border. These instances … exemplify the problems caused by such disputes.

Presently, New Zealand's conflict stems from illegal immigration into its territory, mostly from the Chinese island-province Fujian. … Fujian is situated on China's southern coast, near Taiwan. Many Fujianese immigrants use New Zealand, because of its location, as a stepping-stone to their final goal, the U.S. Their transport is usually a smuggling boat's hold, where living conditions are inadequate and sometimes dangerous, with insufficient food, sanitation, and ventilation. Within the past year, U.S. officials found three Chinese immigrants in a smuggling boat's sealed cargo container, dead from suffocation. … Recently, New Zealand attempted to deal with these aliens by enacting new immigration laws, which hasten the process required to deport them.

The reasons for the Chinese immigrant's journey stems from both "push" and "pull" factors relative to the countries of origin and destination. For example, the Fujianese feel compelled ("pushed") to leave because of the area's low standard of living. The poor wages, bad housing, and lack of political freedom can also be seen as "pull" factors, due to the idea that the Fujianese understand that life would be better in other countries. The U.S. and New Zealand offer much higher wages, a better standard of living, and political freedom. .. These push and pull factors are powerful incentives. … What keeps New Zealand from experiencing an even more profound illegal immigration problem is that the immigrants often do not settle there. …

The border issue between Colombia and its neighbors is another illustration of international conflict. Colombia lies along a corridor from South to Central America. This region has historically been politically unstable, partially due to regional narcotics trafficking, and the wars this engenders. Colombia, itself, is notorious for its export of drugs, especially cocaine. This reputation forces neighboring countries to strengthen patrols over adjoining borders. Recently, Peru deployed additional soldiers to its border with Colombia. Although Peruvian President Fujimori denied any diplomatic problems and stated his troops were there "to guarantee the sovereignty and integrity of Peruvian territory", their mission is both to keep guerrillas *and* drugs out of Peru. … Though understandable, this has in turn, pushed Columbia to respond in kind with more Columbian border troops facing Peru. This brinksmanship seriously depletes resources from these needy countries which might be better spent elsewhere.

Traditionally, these two countries might have been attempting to secure their borders from invading countries, or even seeking to expand their own territories and acquire additional resources. However, Ecuador and Peru are protecting their borders from rogue drug traffickers and guerillas, not Colombia's government. … Neither side is attempting to acquire new territory, but rather to secure and protect that which they already hold. …

29. The author's discussion of "push" and "pull" factors (line 29) most accurately implies that:

A. "pull" factors compel someone to leave, while "push" factors induce someone to come.
B. "pull" factors induce someone to come, while "push" factors also induce someone to come.
C. "push" factors require someone to leave, while "pull" factors also compel someone to leave.
D. "push" factors compel someone to leave, while "pull" factors induce someone to come.

30. Given the information in the passage, if "'artificially' defined" borders (line 3) were eliminated throughout the world, which of the following outcomes would most likely occur?

A. People would naturally immigrate to areas with higher standards of living.
B. Nations would encounter less traditional border strife.
C. Nations would require greater border security measures.
D. People would live more harmoniously.

GO ON TO THE NEXT PAGE.

31. Which of the following assertions does the author support with an example?

- **A.** Transportation methods used by illegal immigrants are sometimes dangerous.
- **B.** Peru and Columbia are seeking to expand their own territories.
- **C.** New Zealand has enacted laws that hasten deportation proceedings.
- **D.** The mission of the Peruvian troops is to keep guerillas and drugs out of Peru.

32. The passage as a whole suggests that in order for a nation to slow the exodus of its inhabitants to other countries, it must:

- **A.** become more attractive to those who are leaving.
- **B.** abandon the traditional methods of guarding borders.
- **C.** respond in some way to the conflicts arising from border disputes.
- **D.** answer the challenges set forth by adjoining countries.

33. If the passage information is correct, what inference is justified by the fact that virtually no immigration from West Berlin to adjoining East Berlin occurred, over the 40 years before the period described?

- **A.** Crossing the heavily guarded borders between West and East Berlin was very dangerous.
- **B.** It was understood that life would be better in East Berlin.
- **C.** The inhabitants of both 'Berlins' were happy to remain where they were.
- **D.** The economic conditions of West Berlin were much more favorable than those of East Berlin.

34. The author implies that which of the following is not one of the reasons that Peruvian President Fujimori deployed soldiers to its borders with Columbia?

- **I.** Fujimori is attempting to keep drugs out of his country.
- **II.** Fujimori fears that Columbia is seeking to expand its territories.
- **III.** Fujimori is probably concerned that Columbia wants to acquire additional resources.

- **A.** I only
- **B.** II only
- **C.** III only
- **D.** II and III only

35. It seems likely that New Zealand may be suffering less from immigration issues than the United States for which of the following reasons:

- **I.** The U.S. offers higher wages than New Zealand.
- **II.** New laws, enacted in New Zealand, allow faster deportation proceedings.
- **III.** Immigrants often do not settle in New Zealand.

- **A.** II only
- **B.** III only
- **C.** II and III only
- **D.** I, II, and III

Passage III (Questions 36-42)

Perhaps the greatest problem with the law of personal injury is its uncertainty about its own purpose—does it exist to compensate victims fully, or to deter careless wrongdoers fully? It must choose, because these two aims are mutually exclusive: tort awards cannot *fully* compensate and *correctly* deter, as long as there are administrative costs involved in obtaining an award. Assume a plaintiff's lawyer charges a 30% contingency fee upon winning a case (or the equivalent flat fee). If the plaintiff is awarded 100% of the damages suffered, she only receives compensation for 70% of her injuries. If she is paid in full, then the defendant is paying 130% of the actual harm caused, and is over-deterred.

In reality, compensation tends toward inadequacy, and not just because of administrative costs. There is no compensation unless the plaintiff proves "negligence", meaning a person may cause any amount of harm, but be excused from paying because she acted "reasonably" rather than carelessly. The hurdle of proving negligence also tends toward inadequate deterrence, because even negligent injurers escape liability if plaintiffs cannot collect convincing proof of negligence.

On the other hand, in a few cases, both compensation and deterrence are exorbitant, especially when a single jury award tries to be both. Consider the following permutation on actual events. A tanker passing through a residential neighborhood leaks acrylonitrile, destroying several homes and poisoning one. Angry residents sue the company for designing its tanker cars negligently. At trial, the company's counsel—a good economist, but a poor lawyer—admits safer tankers were available, but the cost is prohibitive. After extensive cost-benefit analyses, he says, the company found it cheaper just to pay victims for their losses, as it now offers to do. Sound fair? The company would be lucky to escape punitive damages! Remember, these have been applied where judges deemed that even full compensation is inadequate deterrence; generally when the plaintiff's conduct is seen as malicious. However, they may also be awarded when there is "a conscious and deliberate disregard of ... others." ... In this case, the company's cost-benefit analysis is economically "correct" and justifiable: if the new tanker costs more than it saves, it is inefficient. On the other hand, how many juries—or even judges—will see this very analysis as anything but a cold and calculating balancing of profits against the costs of human life? If (arbitrary) punitive damages are granted, plaintiffs emerge overcompensated, and defendants pay out of proportion to harm. Since repeated punitive awards are allowed, the company may be forced to buy the expensive tankers. This is unfortunate, because it results in a waste of resources.

Tort law, gradually realizing the difficulty of proving negligence, has moved towards allowing recovery with ever less proof of negligence. ... Potentially, the most promising development in tort (personal injury) law has been the advent of strict liability, which waives plaintiffs' need to prove the defendant's carelessness in certain instances where the carelessness is obvious, or could have resulted from no factor other than negligence. Unfortunately, the *application* of strict liability is severely constrained by legal doctrine, which limits its application to a small range of "unusually hazardous activities".

Sometimes, the criteria for imposing strict liability seem arbitrary. ... For example, the law permits strict liability only when the expected damage—a product of risk and probable harm—is high. ... Yet it is equally appropriate when the damage is *slight*. Consider the same tanker spilling toxic chemicals along a 200-mile stretch of farmland. Imagine the total cost of decontamination is $1,000,000, but the costs are borne by 150 small farmers. In this scenario, the total damage is high, but comes to only $6,667 per plaintiff. Since just proving negligence may cost more, few will actually sue. The same applies for small harms; there is neither compensation nor deterrence, because plaintiffs bear the loss, and defendants effectively have no incentive to prevent small harms. ...

36. What is the author's response to the standard story about the woman who spills hot McDonald's coffee in her lap, sues and gets several million dollars?

- **A.** This story does not reflect that compensation is usually insufficient.
- **B.** This story is a good example of just the right amount of compensation.
- **C.** This story does not reflect that deterrence is costly.
- **D.** In this story, the woman was malicious.

GO ON TO THE NEXT PAGE.

37. Which of the following assertions is the most effective argument *against* the author's opinion that personal injury law cannot satisfactorily compensate and deter "as long as there are administrative costs involved in obtaining an award" (lines 1-7)?

- **A.** These administrative costs are inconsequential.
- **B.** Attorneys are a necessary part of the judicial system and should be compensated for their work.
- **C.** The administrative costs should be added to the compensation received by the plaintiff.
- **D.** The administrative costs should be subtracted from the compensation received by the plaintiff.

38. The passage indicates that its author would NOT agree with which of the following statements?

- **A.** Tanker companies are a good example of defendants who are under-deterred.
- **B.** Negligence on the part of the defendant is generally not difficult for the plaintiff to prove.
- **C.** The costs associated with suing and defending against suits can be tremendous.
- **D.** In many situations, over-deterrence results in primarily economic ramifications.

39. Assume that since the 9-11 terrorist attacks on the World Trade Center (WTC) buildings, all lawsuits have been settled by the WTC insurance companies, who have now mandated that they will no longer insure any building in the world that is over five stories tall. The author's comments suggest that this situation could reasonably be interpreted as evidence that:

- **A.** the insurance companies were over-deterred.
- **B.** the insurance companies were under-deterred.
- **C.** the plaintiffs were under-compensated.
- **D.** the plaintiffs were overcompensated.

40. Suppose that a study found that police agencies routinely set aside large amounts of money in their yearly budgets, which they expect to pay out in lawsuits against their agency. Which of the following statements is an assumption of the author about the effects of lawsuit awards that would be called into question?

- **A.** Simply proving negligence can be a very costly process.
- **B.** Many people will not sue because the process is too costly.
- **C.** If a plaintiff receives full compensation and administrative costs, the defendant is over-deterred.
- **D.** Depending upon the size of the award, a defendant police agency might not be deterred at all.

41. Which of the following conclusions can justifiably be drawn from the experience of the tanker company's counsel mentioned in the passage?

- **A.** Good economists make for poor attorneys.
- **B.** Costs should never be considered prohibitive where safety is concerned.
- **C.** Toxic materials should not be shipped through residential neighborhoods.
- **D.** Honesty is not always the best policy for an attorney.

42. The author argues that, "Potentially, the most promising development in tort (personal injury) law has been the advent of strict liability" (lines 54-56). These beliefs imply that:

- **A.** the use of strict liability has become increasingly popular for defendants.
- **B.** the uses of strict liability should remain limited in scope.
- **C.** the author approves of waiving the requirement for proof, where carelessness is evident.
- **D.** the author approves of compensation where carelessness is evident.

STOP. IF YOU FINISH BEFORE TIME IS CALLED, CHECK YOUR WORK. YOU MAY GO BACK TO ANY QUESTION IN THIS TEST BOOKLET.

30-MINUTE IN-CLASS EXAM FOR LECTURE 3

Passage I (Questions 43-49)

For most people, gender and gender identity go hand-in-hand. In other words, a female acts like a "woman". The fact that Gender Identity Disorder exists in the DSM IV [the official handbook of psychiatric disorders] as a diagnosis is an admission on the part of psychologists that our society has clearly defined gender roles. These contribute to what it is generally considered "normal". Implied in the name of the disorder is an incongruity between assigned sexual organs and gender identity.

The most comprehensive case study of the disorder is the recently published case history of "Chris," a female patient who identified more closely with males, and stated a wish to become a male. In response, Chris adopted masculine mannerisms such as wearing men's clothes, deepening her voice, sporting a masculine haircut, playing sports against male teams, etc. Chris believed she was born the wrong sex; she was a man inside a woman's body. The incongruity is undisputed; she did everything that was within her means to have her peers recognize her as a man.

It is important to stress when looking for causal factors for Chris' disorder that Chris was in no way confused about her identity. In some ways, this makes Chris an ideal subject for studying the causes of the disorder. She is described as a relatively well-adjusted individual who did well in school, was relatively well liked among her peers, and seemed capable of creating long-term intimate relationships with others. Because the disorder is clearly present (Chris is aware of it herself), basic defects in personality (i.e. maladjustment, inability to make friends) can be more or less ruled out as causes of her uneasiness with her assigned sex. One must then ponder the age-old question of society as the cause.

The notion of gender identity is so heavily dependent on societal norms that, in this case, many psychologists may believe society is the culprit. Similar to the ideas advanced in labeling theory, the mere labeling of some behaviors as "masculine" and others as "feminine" may have created the criteria for this disorder to be labeled as deviant or abnormal. In another culture, where the labeling is different, would Chris have even felt the need to identify herself as distinctly male? Along the same lines of thinking, would the incongruity even be considered a "disorder" in another culture?

Evidence varies regarding the two sides of the issue. Upbringing and/or some biochemical abnormalities could account for the etiology of the disorder, indicating that it is not simply due to societal labels. Indeed, it may seem convincing to argue that Chris strongly identified with males very early in her life, and that this was reflected in her interest in "male" activities (e.g. sports) and unease in using girls' bathrooms and playing on girls' teams. Yet one could also argue that certain activities were labeled male, and Chris molded her interests to include them, so she would be considered "male." It is virtually impossible to unravel which came first, the labeling of her favored activities as "male" or her interest in them. Almost certainly, there is a complex interaction between the two.

Remaining, however, is the question of treatment for Chris. Although she is currently a well-adjusted individual, she acknowledges a desire to change her sexual assignment to match her gender. One might ask if this is really necessary, since she is already well adjusted. Part of what we strive for as psychologically healthy individuals is an acceptance of ourselves in a "natural" state. It is often the case that, through psychotherapy, one learns that one may not necessarily have to change oneself as much as one's perception of self. ... The effect of a physical sex reassignment operation on Chris' happiness cannot be foretold with complete certainty.

In conclusion, it is interesting to note that Chris' desire to have the anatomy of a man is considered part of a "disorder," while a small-breasted woman's desire for breast implants would usually be construed as a desire to increase her femininity and not be labeled as such. Perhaps Chris is somewhere on a male-female continuum and, like the woman who desires the breast augmentation, is pushing herself toward the closer end of the spectrum into our neatly constructed gender dichotomy.

43. The passage suggests that its author would probably *disagree* with which of the following statements?

- **A.** It is possible that Chris participated in "male" activities in order to be considered male.
- **B.** It is possible that Chris naturally participated in "male" activities.
- **C.** Chris was not confused about her identity.
- **D.** Most cultures have clearly defined gender roles.

GO ON TO THE NEXT PAGE.

44. Implicit in the passage is the assumption that:

I. one should be happy in one's "natural" state.
II. one can be well-adjusted, yet unhappy with one's "natural" state.
III. one's perception of self is most important.

A. I only
B. II only
C. III only
D. I and III only

45. The author of the passage would be most likely to agree with which of the following ideas expressed by other psychologists?

A. A DSM IV 'disorder' may not actually be a disorder at all.
B. The DSM IV is a poor descriptor of abnormal behavior and desires since it is easily influenced by societal norms.
C. Some DSM IV 'disorders' are simply an attempt to characterize socially abnormal behavior and desires.
D. Behavior and desires must fall within the parameters of the DSM IV to be considered normal by society.

46. The author hints that the fact that Chris is well-adjusted indicates that her "uneasiness with her assigned sex" (lines 30-31):

A. is a problem which should be overcome through psychiatry.
B. is due to the culture she lives in.
C. can be overcome through surgery.
D. is a basic personality defect.

47. Suppose it is discovered that prescription medication allows Chris to become somewhat more comfortable with her "natural" state. Does this discovery support the author's argument?

A. Yes; it confirms it.
B. No; it does not affect it.
C. No; it weakens it.
D. No; it disproves it.

48. The author admits that, "The effect of a physical sex reassignment operation on Chris' happiness cannot be foretold with complete certainty" (lines 67-69). The author most likely believes that:

I. it is just as likely that psychotherapy would help Chris to change her perception of self.
II. in our society, in the body of a woman, Chris will not be happy.
III. a "sex reassignment operation" would make Chris happier.

A. I only
B. II only
C. III only
D. II and III only

49. The author's attitude toward "our" societal norms is most accurately described as:

A. favorable.
B. neutral.
C. distrustful.
D. disapproving.

GO ON TO THE NEXT PAGE.

Passage II (Questions 50–56)

Human values – these 'ends', which are generally viewed as "desirable" – are not static. They change according to certain factors surrounding the people who create them. One of the more significant of these factors is technology. Due to this, people must continually question the reasons for the use of technology in their own surroundings; in their "information ecologies".

Just as "freedom of speech" is not clearly defined, the purpose of some technology is ambiguous. To decide whether a certain technology is a good fit for a certain "ecology," one must use critical analysis. [Analysts] Nardi and O'Day create a three-pronged format for analyzing technology's worth. ... These methods include working from core values, paying attention, and asking strategic questions.

Making sure that technology fits with a person's or group's "core values" – its essential function/purpose – is the first step in examining its use. If technology doesn't assist in promoting these values, then it cannot be considered useful. For example, a for-profit business' core value is profit maximization; any technology which does not increase revenues or reduce costs should be viewed with suspicion. ... Paying attention keeps people from taking a technology for granted. If the technology is taken for granted, the actual purpose for using it can become clouded. Finally, asking strategic and open-ended questions fosters a discussion about all uses for the technology. Asking who, what, why, where, and how questions is the best method in analyzing the use for technology in an information ecology. If, by examining the situation, it becomes evident that the existing technology is not useful, then the person or group that expects to use it should have direct input into finding a replacement. This ensures that technology will be used for the right purposes.

Technology may force people away from their own values. This idea, presented by theorists Neil Postman and Jacques Ellul, creates a disturbing image of a world where technology itself becomes the center of a society without values. One example is "information glut", a term coined by Postman. He states that the rapid flow of information allowed by automation and Internet/email provides vast amounts of data to many users quickly and simultaneously, but doesn't allow humans to understand its meaning fast enough to keep up. He goes on to say that much of the information we receive does not pertain to us, so it is at best an irrelevant distraction rather than the useful news it was intended as. This may create a problem when, for example, images of violence in the Middle East do not affect us because they are "here today, gone tomorrow" and are too far away to influence us. However, this view is somewhat one-sided. It is argued that too much information is bad, yet it can also be argued that more information is good. By giving people more information, they have a wider basis from which to make decisions and form opinions. No one can argue that giving additional and different perspectives is inherently bad. Yet, Postman does have a valid point that the information is useless unless it is thoroughly examined. Simply put, the challenge is not necessarily to reduce the amount of information available, but to allow it to be better tailored to individual recipients' needs. Thus, we as a society must consciously attempt to not only obtain information, but to analyze it for meaning and relate this back to our own values. ...

50. The author of the passage would be most likely to agree with which of the following ideas expressed by other technology theorists?

- **A.** Soliciting further ideas and diverse ideas is not always wrong.
- **B.** Obtaining additional and differing perspectives is always beneficial.
- **C.** For information to be of value, it must challenge our human values.
- **D.** A great deal of the information which we receive is useless because it does not pertain to us.

51. The passage argument suggests that information recipients might benefit from:

- **I.** a more rapid flow of information.
- **II.** thoroughly examining their information.
- **III.** limiting non-pertinent information.

- **A.** I only
- **B.** II only
- **C.** III only
- **D.** II and III only

52. In describing Neil Postman's "information glut", the author uses the example of "images of violence in the Middle East [that] do not affect us" (lines 48). The author's point is that:

A. the Middle East is at best an irrelevant distraction.
B. the Middle East is poorly understood by us.
C. these images should affect us.
D. these images should not be so easily accessible.

53. If the following statements are true, which would most weaken the argument of the author?

A. The degree to which technology determines human values is questionable.
B. Information must be analyzed for its relevance to human values.
C. Human values are unchanging.
D. Human values are culturally dependent.

54. The author argues, "more information is good" (line 52). Unlike Neil Postman, the author does not consider which of the following to be a factor that might limit the usefulness of information?

A. Space
B. Distance
C. Frequency
D. Time

55. According to the passage, one drawback of technology is that it can:

A. supersede all other values.
B. be used to destroy human values.
C. subtly change people's values.
D. become an irrelevant distraction.

56. Suppose it could be established that technology is most efficient when it performs its function unobtrusively; without being noticed. The author of the passage would be most likely to respond to this information by:

A. suggesting that this determination ratifies his thesis.
B. proposing that we must still analyze the information for meaning.
C. asserting that efficiency is usually degraded when a technology is taken for granted.
D. explaining that we must still remain aware of the technology and its intended purpose.

GO ON TO THE NEXT PAGE.

Passage III (Questions 57–63)

From its very beginning, … the New York City Opera production of "Mephistopheles" deserves high marks for visual excellence. It begins with an audiovisual show featuring stars, religious images projected onto swirling mist, and very, very loud brass winds, intended for drama and only slightly corny.

The scene then shifts to Hell, where a naked, disheveled Mephisto, singing from his broken throne, sarcastically apologizes for not being up to Heavenly standards of singing, providing the proof that harmony is still a longer way off in some places than in others. In the director's vision, the characterization of Mephisto is akin to Milton's rebellious, but somewhat sympathetic anti-hero, a dissident angel who dares to fight a vastly superior power to preserve his vision of the world. Accordingly, this production features a Mephisto who is flippant but clever, blasphemous but thought provoking, and possessed of both sympathy and contempt for human weaknesses. He strikes a balance between his boldness and his cowering before Heaven.

A chorus presents the essential plot of Faust, reduced from its several incarnations. When the angels point out the mortal Dr. Faust, as God's incorruptible servant on earth, the devil Mephisto promises to turn him from God through temptation. Thus, this version presents temptation as essentially a wager, or struggle, between God and the Devil (which, at one time, was a remarkably blasphemous notion, as it contradicted the dogma that God is all-powerful over evil). As the divine host departs, Mephisto regains his mocking manner, singing, "It's nice to see the Eternal Father talking with the Devil—in such a human way!"

Mephisto tempts Faust in the middle of a country fair thronged with revelers, which is meant to symbolize the worldly pleasures. This symbolism hearkens back to the ancient morality play *Vanity Fair*, which also featured a bazaar extolling sins. Mephisto, garbed in virtue as a gray robed beggar monk, finally announces himself to be Mephisto. In a good aria, much of which is delivered while dancing or rolling on the ground, the devil again introduces himself sympathetically as God's constructive critic, one who "thinks of evil/ but always achieves the good," singing menacingly of how he wages an eternal dissent against God:

"Light has usurped my power,
seized my scepter in rebellion;
I hurl forth this single syllable—NO!"

In this version of Faust, it is Faust who seizes the devil's bargain: Mephisto must furnish him with a single moment so lovely it deserves to last forever.

In Scene II, Faust is transformed into a younger man, who courts the young woman of his dreams, a commoner named Margaret. At this point, the play devolves into stock characters and slapstick. Faust and Margaret sing very forgettable arias about the supremacy of feeling over reason, a theme which is not really congruent with the Faust myth. The shallowness of the libretto's throwaway lyrics is compounded by Margaret's emotionless singing.

Those who read the book know the next scene as the Witch's Sabbath on Walpurgis Night, though the libretto itself offers little explanation for the abrupt change of scene. Mephistopheles now appears as the leader of hedonistic sinners, calling them with the aria "Come on, onward, onward". Again, the portrayal of his character is less evil than rebellious and hedonic, he recognizes the power of mankind's pursuit of earthly pleasures, singing, "Here is the world, round and empty" while holding the globe. It is unclear whether it is his effort, or human nature itself, which is responsible for sin. At one point, he laments *mankind*'s cruelty and cunning, concluding, "How I laugh when I think what's in store for them!/Dance on; the world is lost." …

57. Assume that several others who had attended the same opera were interviewed. If they remarked that Margaret sang with tremendous passion, these remarks would weaken the passage assertion that:

A. Mephisto had rendered her irresistible to Faust.
B. the lyrics which she sang were "throwaway".
C. the young woman of Faust's dreams sang without emotion.
D. the young woman of Faust's dreams was a commoner.

GO ON TO THE NEXT PAGE.

58. On the basis of the passage, it is reasonable to conclude that:

A. Faust lost his soul to the devil.
B. the "country fair thronged with revelers" was not in the book.
C. the operatic interpretation differed from the book.
D. the author did not enjoy the performance.

59. According to the passage, the author felt that the New York City Opera production of "Mephistopheles":

A. was plagued with a poor characterization of Mephisto.
B. suffered from noticeable weaknesses beginning in Scene II.
C. was enhanced by Dr. Faust's singing.
D. could have been improved in Scene III.

60. According to the passage, the author seems to have most enjoyed:

A. the music of the opera.
B. the singing of the opera .
C. the plot of the opera.
D. the images of the opera.

61. Which of the following does the author suggest was a component of the original "Faust myth" (lines 55-56)?

I. Reason triumphing over feeling
II. A more evil Mephisto
III. A more powerful God

A. I only
B. II only
C. III only
D. II and III only

62. According to the passage, through what primary means is the fundamental plot transmitted to the audience?

A. Via visual imagery
B. Via Faust's musings
C. Via the chorus
D. Via Mephisto

63. Regarding the devil's bargain with Faust, the passage strongly implies that:

A. it is Faust who got the better deal.
B. it is the devil who got the better deal.
C. in other versions, the bargain was with Margaret.
D. in other versions, it is Faust who does the bargaining.

STOP. IF YOU FINISH BEFORE TIME IS CALLED, CHECK YOUR WORK. YOU MAY GO BACK TO ANY QUESTION IN THIS TEST BOOKLET.

ANSWERS & EXPLANATIONS
FOR
IN-CLASS EXAMS

ANSWERS TO THE IN-CLASS EXAMS

Exam 1	Exam 2	Exam 3
1. C	22. D	43. D
2. B	23. A	44. B
3. D	24. C	45. C
4. C	25. A	46. B
5. A	26. B	47. B
6. C	27. A	48. D
7. B	28. C	49. D
8. B	29. D	50. A
9. A	30. B	51. B
10. D	31. A	52. C
11. C	32. A	53. C
12. A	33. D	54. D
13. C	34. B	55. C
14. B	35. C	56. D
15. B	36. A	57. C
16. C	37. B	58. C
17. C	38. B	59. B
18. A	39. A	60. D
19. B	40. C	61. A
20. D	41. D	62. C
21. B	42. C	63. D

MCAT VERBAL REASONING AND MATH

Raw Score	Estimated Scaled Score
23	15
22	14
21	13
19–20	12
18	11
16-17	10
15	9
13-14	8
12	7
10-11	6
9	5
7-8	4

EXPLANATIONS FOR 30-MINUTE IN-CLASS EXAM 1

Passage I (Questions 1–7)

1. The word tar (line 53) is used in the sense of:

"Some Marxist theorists have speculated that Marx would tar both Kant and Hume as "bourgeois" philosophers" (lines 52-53).

A. asphalt.
WRONG: This is not the intended sense, or meaning, of the word. "Tar" is not used literally.

B. suggest.
WRONG: This is not the intended sense, or meaning, of the word. This is not strong enough. *

C. label.
CORRECT: This is the intended sense, or meaning, of the word. The reference harkens to a literal "tarring and feathering" of individuals for the purpose of ridiculing and labeling them as collaborators. However, this knowledge was not necessary in order to answer the question. In the context of the passage, the author clearly theorizes that Marx would strong disagree with Hume and Kant. Substituting the word "label" for "tar" is appropriate.

D. stick.
WRONG: This is not the intended sense, or meaning, of the word. This word is not specific enough.

2. According to the passage, Marx would have disagreed with Kant and Hume over which of the following ideas?

The correct answer requires the embodiment of an idea which is 1. shared by Kant and Hume, and 2. that Marx would have disagreed with.

A. What each man feels within himself is the standard of sentiment.
WRONG: This is not an idea that Kant and Hume share. This is Hume's idea only (lines 14-15).

B. There is uniformity in people's reason or emotion.
CORRECT: First, this is an idea that Kant and Hume share; "To make both of [Kant's and Hume's] constructions possible, there is a notion of some kind of uniformity in people's reason or emotion in both works" (lines 43-44). Second, according to the passage, this is clearly an idea with which Marx would have disagreed. "Based on these writings, Marx would probably see any system that sought out a universal theory of morality as ignoring the opposing economic classes in society ..." (lines 65-67).

C. The interests of the capitalists and the workers are one and the same.
WRONG: This is not clearly an idea that Kant and Hume share. It is outside the scope of information in the passage.

D. Morality is dependent upon the class struggle.
WRONG: This is not an idea from the passage.

3. Assume that a universal principle of morality can be proven to exist. Which of the following hypotheses does this assumption suggest?

A. The author is correct; despite their genesis, it is not surprising that Kant and Hume constructed similar systems.
WRONG: This hypothesis is not suggested by the assumption in the question. "Considering that they began their searches with seemingly irreconcilable ideas of where to look, the similarity in the moral systems they constructed is surprising" (lines 3-6).

B. The author is correct; Marx, Hume, and Kant all constructed similar systems.
WRONG: This hypothesis is not suggested by the assumption in the question. The author never suggests that all three of these people did construct similar systems.

C. The author is incorrect; Marx, Hume, and Kant did not all construct similar systems.
WRONG: This hypothesis is not suggested by the assumption in the question. The author never suggests that *all three* of these people did construct similar systems.

D. The author is incorrect; despite their genesis, it is not surprising that Kant and Hume constructed similar systems.
CORRECT: This hypothesis is suggested by the assumption in the question. "Considering that they *began their searches with seemingly irreconcilable ideas* of where to look, the similarity in the moral systems they constructed *is surprising*" (lines 3-6). It is not so surprising once given the assumption that a universal principle of morality can be proven to exist. Hume and Kant simply arrived at this "Truth" by different paths.

4. According to the author, in creating his moral system, Hume equated:

A. circumstances with disinterested comparison.
WRONG: There is no support for this answer in the passage.

B. circumstances with absolutism.
WRONG: The two are diametrically opposed in the passage and not equated by Hume.

C. practicality with stirring people to action.
CORRECT: "Hume decided at the outset that *a moral system must be practical*, ... and since only sentiment (emotion) is capable of *stirring people to action*, the practical study of morality should be concerned with sentiment" (lines 7-9).

D. practicality with reason.
WRONG: Hume was not much impressed with reason. He focused on sentiment.

5. Based upon passage information, Marx most likely believed that:

A. there is no universal theory of morality.
CORRECT: This is Marx's most likely belief. "Marx would probably see any system that sought out a universal theory of morality as ignoring the opposing economic classes in society ...[which he felt were involved in a class struggle]" (lines 65-67).

B. philosophers were part of the bourgeois.
WRONG: Based upon passage information, this is not Marx's most likely belief. This is a vast generalization without any real support. We have no way of knowing what Marx thought about philosophers in general.

C. the workers could be easily fooled.
WRONG: Based upon passage information, this is not Marx's most likely belief. This is a vast generalization without any real support.

D. Kant's book supported capitalist exploitation.
WRONG: Based upon passage information, this is not Marx's most likely belief. There is no passage information that would tell us whether Kant's book had ever been read by Marx, whether it was written before Marx's time, or written well afterwards. This is outside the scope of the available passage information.

6. Based upon passage information, Kant's system of "moral feedback" (line 25) differed from Hume's in that it might result in a situation wherein:

A. one realized that his action might be 'right' so long as it didn't become a universal law of nature.
WRONG: Hume did not consider "universal laws of nature". Further, and more importantly, the idea of a "universal law of nature" (lines 31-32) was all important to Kant.

B. one realized that a universal law of nature was unnecessary in determining duty.
WRONG: It was Hume who did not consider "universal laws of nature". Further, and more importantly, the idea of a "universal law of nature" (lines 31-32) was all important to Kant.

C. one realized that his action might be 'wrong' even though it created agreeable sentiments in others, as well as in oneself.
CORRECT: Unlike Hume, Kant's system relied upon reason and logic [think Vulcan]; "before one does anything, one should forget one's own motives for a moment, and ask if he would want everyone to do as he does. If the answer is no, then his subjective desire is different from his objective assessment, and the action is contrary to duty" (lines 32-36). Sentiment was not a factor for Kant. A "subjective desire" that might, in the short term please himself and others, might in the long term not be best established as a "universal law of nature" (lines 31-32).

D. one realized that his action might be illogical if sentiment was not further considered.
WRONG: Sentiment was not a factor for Kant.

7. According to the passage, Hume's ideas evolved to the point where he:

Note that the correct answer *requires* that you know and understand where Hume's ideas stood when he began formulating them. "Hume's ideas *evolved* ...". You must consider the paragraph beginning, "Hume decided at the outset ..." (lines 7-15).

A. realized that reason was an inseparable part of a universal system of morality.
WRONG: Hume did not think much of reason (lines 7-9), but based his ideas upon sentiment (emotion).

B. was considering the sentiments of others as well as himself.
CORRECT: As indicated by the question, Hume's ideas *evolved*. They changed from the time he first began formulating them to the end. At first, Hume says "What each man feels within himself is the standard of sentiment" (lines 14-15); others are not considered. Later, "Hume says that moral actions are those that create agreeable sentiments in *others*, as well as in oneself" (lines 37-38).

C. chose to essentially agree with Kant on a universal system of morality.
WRONG: This is not correct. Even in the end, the two differed in their reliance on reason/logic as opposed to sentiment/emotion.

D. decided that sentiment without action was a necessary component of his morality system.
WRONG: Hume chose sentiment specifically because he felt that it was only sentiment which elicited action. "Hume decided at the outset that a moral system must be practical, ... and since only sentiment (emotion) is capable of stirring people to action ..." (lines 7-10).

Passage II (Questions 8-14)

8. Which of the following statements, if true, would most directly *challenge* the principles of Martin Luther?

A. The Bible's Old Testament refers to a period before the birth of Jesus.
WRONG: This does not most directly challenge the principles of Martin Luther. This answer is way outside of the scope of the passage.

B. The Bible alone contains only a small part of what Jesus intended for his followers.
CORRECT: This most directly challenges the principles of Martin Luther. This is within the scope of the passage and relates directly to information provided in the passage on the beliefs of Martin Luther; "For Luther, salvation lies in faith emanating from personal understanding of Biblical teachings" (lines 56-57).

C. The German "barbarians" who sacked Rome had been previously converted.
WRONG: This does not most directly challenge the principles of Martin Luther. Whether they had been converted or not is irrelevant. Either way, they would have been considered "living tools" of God by both Luther and Augustine.

D. Augustine's understanding of salvation granted through the mercy of Christ was flawed.
WRONG: This does not most directly challenge the principles of Martin Luther. It has little bearing on the principles of Martin Luther. If anything, this information would *strengthen* Luther's ideas and principles.

9. Some theologians believe that killing and violence are acceptable when used in self-defense. An appropriate clarification of the passage would be the stipulation that:

A. both Luther and Augustine would have disagreed with this belief.
CORRECT: This stipulation would be an appropriate clarification. "Both of these theologians would hold that … salvation, … should be all Christians' only concern in life" (lines 52-55). There is no evidence in the passage that either theologian made exceptions to their explanations of Christ's teachings wherein "killing and violence" are acceptable.

B. both Luther and Augustine would have agreed with this belief.
WRONG: This stipulation would not be an appropriate clarification. There is no evidence in the passage that either theologian made exceptions to their explanations of Christ's teachings wherein "killing and violence" are acceptable.

C. only Augustine might have agreed with this belief.
WRONG: This stipulation would not be an appropriate clarification. "In Augustine's understanding, [God] sends hardship and death to test men's faith. The true Augustinian Christian would maintain his faith [and presumably his passivity] through any ordeal, and even if his body perishes, God will save him for his conviction" (lines 59-64).

D. only Luther might have agreed with this belief.
WRONG: This stipulation would not be an appropriate clarification. See above.

10. If the information in lines 38-55 is correct, one could most reasonably conclude that, compared to Luther, Augustine was:

A. much more reasonably inclined.
WRONG: This is not a reasonable conclusion. This is a fuzzy value judgment based solely upon personal opinion. Unless the passage itself provided direct evidence regarding the "reasonableness" of someone's inclinations or ideas, this type of answer would be a poor choice.

B. more prepared to define God's will.
WRONG: This is not a reasonable conclusion. There is no evidence in support of this. What does "more prepared" mean? This answer approaches the vagueness of Answer A.

C. less eager to send people to eternal torment.
WRONG: This is not a reasonable conclusion. This answer is attractive. Particularly given the last sentence of the quote, "Augustine never considers the individual fate of these living tools, but Luther maintains that … they go to eternal torment" (lines 49-51). However, neither of the theologians is "sending people to eternal torment". According to the passage and the information it is God who decides who is 'sent' to hell, not Luther or Augustine.

D. less willing to announce God's final judgment on those who had sinned.
CORRECT: This is the only reasonable conclusion based on the information. "Augustine never considers the individual fate of these [barbarians], but Luther maintains that ... they go to eternal torment" (lines 49-51). This answer choice more correctly defines the role that Luther or Augustine might play as theologians in relation to God. The theologians can only "announce" what God has decided; if they think they know. They themselves do not have the power to "send" anyone to heaven or hell.

11. The author's attitude toward the theories of Augustine and Luther in the passage is most accurately described as:

This type of question must usually be gleaned from the overall impression given by the author as the passage is read. There is no 'going back' to the passage. The reader must ask if there were any derogatory, sarcastic, praiseworthy, or other type of information or words used that provide clues to the author's attitude. If the passage did not try to persuade or argue, then it is probably neutral.

A. disapproving.
WRONG: This is not the most accurate description of the author's attitude.

B. mistrustful.
WRONG: This is not the most accurate description of the author's attitude.

C. neutral.
CORRECT: This is the most accurate description of the author's attitude. This type of passage is not persuasive, but purely informative.

D. favorable.
WRONG: This is not the most accurate description of the author's attitude.

12. What is the meaning of the phrase; "The true Augustinian Christian would maintain his faith through any ordeal, and even if his body perishes, God will save him for his conviction" (lines 61-64)?

Unfortunately, passages and sentences are not always provided in their clearest form. The tortured syntax of this sentence begs for clarification. Particularly where the word "conviction" is used in its less common form; 'belief'. We are used to thinking of a prosecuting attorney seeking a criminal's 'conviction'.

A. This Christian would end up in heaven because of his beliefs.
CORRECT: This is the meaning of the phrase. Restated in other words, 'The true Christian would maintain his faith through any ordeal, and even if he physically died, God will save him *because of his belief*.

B. God would save this Christian for judgment at the end of the Christian's ordeal.
WRONG: This is not the meaning of the phrase. This answer mistakenly equates "for his conviction" with 'for his judgment'.

C. God would save this Christian from his ordeal and judge him.
WRONG: This is not the meaning of the phrase. First, this Christian will not be saved from his ordeal. It is clear that God is not going to save this Christian from his 'earthly' ordeal. "[E]ven if his body perishes ..." means the Christian *will* probably die. Secondly, this answer mistakenly equates "for his conviction" with 'for his judgment'.

D. It was not necessary for this Christian to die for him to be convicted.
WRONG: This is not the meaning of the phrase. First, this Christian will not be saved from his ordeal. It is clear that God is not going to save this Christian from his 'earthly' ordeal. "[E]ven if his body perishes ..." means the Christian *will* probably die. Secondly, though this answer is not completely clear, it seems to mistakenly equate "for his conviction" with "finding him guilty".

13. The author's primary purpose in the passage is apparently:

These questions are less value judgments on the word "primary" than they are accuracy questions. On these types of questions the "primary purpose" usually means the answer choice which accurately restates passage information. Generally, three are inaccurate and one is accurate. However, if two answer choices are accurate, one will clearly be more all encompassing (primary) than the other.

A. to clarify the differences between the ways in which the early Catholics and Protestants dealt with persecution.
WRONG: This is not the author's primary purpose. There seem to be *no* differences in the passage regarding the way in which the early Catholics and Protestants dealt with persecution. The only differences were in their understandings of how to achieve salvation and pronouncing God's judgment on others.

B. to justify the persecution of early Christians by secular governments.
WRONG: This is not the author's primary purpose. The only justification of persecution in the passage comes indirectly from Augustine's writings "that His inscrutable will is always just" (lines 40-41), and that "God ... sends hardship and death to test men's faith" (lines 60-61).

C. to consider the similarities between the ways in which the early Catholics and Protestants dealt with persecution.
CORRECT: This is the author's primary purpose. A topic sentence in the first paragraph is, "Later Christian theorists followed this example, often urging peaceful coexistence even with governments which violated every precept of Christian teachings" (lines 7-10). The author then goes on to support this idea with Augustine and Luther by considering their "similarities". Though Luther's ideas predominate in the passage, "St. Augustine would have agreed" (line 38), and "Both of these theologians would hold that ..." (line 52), etc.

D. to question the passive practices of the early Catholics and Protestants when faced with persecution.
WRONG: This is not the author's primary purpose. The author is not questioning the practices, but explaining the reasoning behind them in a rather objective manner.

14. What is the most serious apparent weakness of the information described?

These types of questions are not necessarily the value judgments that they might first seem to be. At least two of the answer choices will inaccurately restate conclusions or passage information. Of the remaining two, one will obviously weaken the passage, while the other may even strengthen it.

A. While implying that Christians may coexist with a secular government, it differentiates between Catholics and Protestants.
WRONG: This is not the most serious apparent weakness of the information described. What little differentiation there is *strengthens* the passage by supporting the idea that "Later Christian theorists ..." (i.e. supporting the topic sentence).

B. While implying representation of Augustine and Luther, its conclusions are based primarily on information according to Luther.
CORRECT: This is the most serious apparent weakness of the information described. The author chooses the Catholic pillar St. Augustine and the Protestant representative Martin Luther to stand for later "Christian theorists". However, the second paragraph merely paraphrases the two of them. The large third paragraph is all Luther. Finally at line 38, "St. Augustine would have agreed" but by the next sentence we are back to Luther. The passage does not even come close to equally representing the two Christian theorists and is therefore weakened when attempting to use both men's theories in support of the topic.

C. While implying representation of all Christian theorists, only Augustine and Luther are mentioned.
WRONG: This is not the most serious apparent weakness of the information described. If you felt that there were other Christian theorists from that time you would be bringing in outside information. For instance, Fundamentalists, Mormons, and other sects did not exist at that time, for one thing.

D. While implying agreement between Augustine and Luther, their attitudes were clearly opposed.
WRONG: This is not the most serious apparent weakness of the information described. This is inaccurate and there is no support for this answer.

Passage III (Questions 15–21)

15. The author probably mentions that Mallon was "never 'tried' in any sense" (line 5) in order:

A. to demonstrate the power of the wealthy at that time.
WRONG: This is not the probable reason for the author's mentioning. Though the privilege of the wealthy is an underlying theme of the passage, there is no evidence that the "wealthy", nor their power, had anything to do with Mallon's confinement.

B. to provide a comparison with people who have actually committed a crime.
CORRECT: This is the most probable reason for the author's mentioning. This answer supports the fact that the author obviously feels that Mallon's treatment was unjust. This is evidenced in the first and last sentences of the passage: "Of all the bizarre and melancholy fates that could befall an otherwise ordinary person, Mary Mallon's has to be among the most sad and peculiar.", and "But she was the only one kept isolated for years, a result as much of prejudice toward the Irish and noncompliant women as of a public health threat.". This is the best answer.

C. to illustrate the persistence of Soper's investigations.
WRONG: This is not the probable reason for the author's mentioning. There is no evidence that Soper's investigations were 'persistent' in the first place.

D. to support the claim that she deserved at least a hearing.
WRONG: This is not the probable reason for the author's mentioning. There is no "claim" by the author that she "deserved at least a hearing".

16. According to the passage, the first two investigators were unable to find the cause of the outbreak (lines 19-20). The information presented on typhoid makes which of the following explanations most plausible?

A. They focused too closely on the "soft clams" that Soper later discredited.
WRONG: This explanation is not the most plausible. There is no evidence or information that the investigators focused on "soft clams" at all.

B. Typhoid is not really passed through contaminated food or water.
WRONG: This explanation is not the most plausible. It seems that typhoid *is* passed through contaminated food or water.

C. They never considered that typhoid could be carried by a healthy person.
CORRECT: This explanation is the most plausible. We know that Soper "was the first to identify a healthy typhoid carrier in the United States" (lines 32-33). Also, Soper's investigation and discovery occurred after the two investigators came to the Warren house. Therefore, this answer is the most likely.

D. By this time, Mallon was no longer employed by the Warren family.
WRONG: This explanation is not the most plausible. There is no timeframe given regarding the "two investigators", thus this is purely speculative. Further, since, Soper was the first person to identify a healthy typhoid carrier in the United States, it is unlikely that the two investigators would have suspected Mallon had she been even directly interviewed by them..

17. The author's argument that Mallon's isolation was "a result as much of prejudice ... as of a public health threat" (lines 64-65) is most *weakened* by which idea in the passage?

Note that the correct answer must satisfy two criterions. It must 1). be an idea in the passage, and 2). weaken the author's argument.

A. Mallon's primary occupation was as a cook.
WRONG: The author's argument is not weakened by this idea, *to the extent that Answer C weakens it*. First, though it does seem that this answer is promoted in the passage, it is an assumption. If Answer C were not available this might be a good choice. However, it is not the fact that Mallon was a cook which doomed her, but it was the fact that she "failed to comply with the health inspector's requirements" (lines 58-59), which presumably would have precluded her from preparing meals. C is the better answer.

B. Mallon did not believe that she was a carrier of the disease.
WRONG: The author's argument is not weakened by this idea, *to the extent that Answer C weakens it*. If Answer C were not available this might be a good choice. However, similarly to Answer A, it is not the fact that Mallon did not believe that she was a carrier which doomed her, but it was the fact that she "failed to comply with the health inspector's requirements" (lines 58-59). C is the better answer.

C. Mallon would not abide by the health inspector's requirements.
CORRECT: The author's argument is weakened by this idea. Ultimate, this is the explicitly provided reason for Mallon's reincarceration. "After a short period of freedom in which *Mallon failed to comply with the health inspector's requirements*, she was eventually sent back to North Brother Island, where she lived the rest of her life, alone in a one-room cottage" (lines 58-61).

D. Mallon actually was the source of the typhoid outbreak in the Warren home.
WRONG: The author's argument is not weakened by this idea, *to the extent that Answer C weakens it*. Had Mallon complied with the "health inspector's requirements" (line 59), she might well not have been reincarcerated.

18. Which of the following statements is the most reasonable conclusion that can be drawn from the author's description of the typhoid outbreak in the house at Oyster Bay?

A. The Warren family did not hire Soper.
CORRECT: This is a reasonable conclusion. "Worried they wouldn't be able to rent the house unless they figured out the source of the disease, *the owners ... hired George Soper ...*" (lines 20-21).

B. The two investigators were hired by the Warrens.
WRONG: This is not a reasonable conclusion. It is *not* certain who it was who hired the two investigators. Why not the owners, or the city health department, for instance?

C. The Warren family hired Soper.
WRONG: This is not a reasonable conclusion. "Worried they wouldn't be able to rent the house unless they figured out the source of the disease, *the owners ... hired George Soper ...*" (lines 20-23).

D. The owners were anxious to sell the house.
WRONG: This is not a reasonable conclusion. The owners were worried that they wouldn't be able to *rent* the house. "Worried they wouldn't be able to rent the house unless they figured out the source of the disease, *the owners ... hired George Soper ...*" (lines 20-23).

19. Passage information indicates that which of the following statements must be true?

A. Mallon had probably not infected anyone prior to the Warren family.
WRONG: There is no information that this *must* be true.

B. Mallon was almost certainly not washing her hands prior to preparing the Warren's meals.
CORRECT: We know that Mallon's feces "did indeed show high concentrations of typhoid bacilli" (lines 46-47). Therefore, there are only a few unlikely alternatives to the idea that the bacilli were being trans-

ferred from the feces to Mallon's hands, to the food she prepared, and then to those becoming sick. None of the grotesque alternatives is really plausible.

C. Being labeled 'Typhoid Mary' by the press was the primary reason for her confinement.
WRONG: There is no information that this *must* be true.

D. The health inspector was doubtless prejudiced toward the Irish.
WRONG: There is no information that this *must* be true.

20. According to passage information, Mallon worked for the Warren family for approximately:

A. two weeks.
WRONG: See explanation for Answer D.

B. three weeks.
WRONG: See explanation for Answer D.

C. five weeks.
WRONG: See explanation for Answer D.

D. six weeks.
CORRECT: According to Soper's own words describing Mallon's tenure in Oyster Bay, which are quoted in the passage, "It was found that the family had changed cooks about *three weeks before* the typhoid epidemic broke out ... She remained with the family only a short time, *leaving about three weeks after* the outbreak occurred" (lines 26-29).

21. The contention that in "1938 when [Mallon] died, ... there were 237 other typhoid carriers living under city health department observation. But she was the only one kept isolated for years" (lines 61-65), can most justifiably be interpreted as support for the idea that:

A. Mallon was unfairly treated by the city health department.
WRONG: This idea is not justifiably supported by the contention. "Fairness" requires a comparative analysis of some sort. Mallon was the "first" of her kind to be identified. The contention regards 1938, decades after Mallon's discovery. Her treatment cannot be compared to the 237 other typhoid carriers.

B. Mallon's isolation might have stemmed from the health department's early ignorance of the disease.
CORRECT: This idea is justifiably supported by the contention. This answer might be gleaned through process of elimination also. However, the key "*early* ignorance" is a giveaway given that the contention regards 1938, decades after Mallon's discovery and isolation.

C. The "other 237 typhoid carriers were all kept isolated at one time or another.
WRONG: This idea is not justifiably supported by the contention. Though this is tempting because the contention provides that "she was the only one *kept isolated for years*", this is not a strong interpretation of the contention. It is entirely possible that at least one of the 237 "other" carriers was never isolated at all, negating this answer.

D. The "other" 237 typhoid carriers were much like Mallon.
WRONG: This idea is not justifiably supported by the contention. This is very weak. For one thing, the contention regards 1938, decades after Mallon's discovery. Further, it can be presumed, based upon the passage, that those who were free had at least agreed and cooperated in not cooking and serving food; unlike Mallon.

ANSWERS & EXPLANATIONS FOR 30-MINUTE IN-CLASS EXAM 2

Passage I (Questions 22–28)

22. According to the passage information, which of the following would be most likely, if a person, who was talking to you, attempted to make their voice sound unusually low?

A. You might think that they were lying.
WRONG: This would not be most likely. The key is "*unusually* low". Though a speaker who was lying, might be aware that in order to successfully lie, or "suggest credibility, the voice's tone should be pitched as low as is naturally possible" (lines 4-5), this is a much more 'tortured' answer choice. In other words, in contrast to the straightforward aspect of Answer D, this answer requires several assumptions that are outside the scope of the question.

B. They could be irritated with you.
WRONG: This would not be most likely. First, it is lack of "intelligibility" that is irritating. Second, it is the listener who is irritated, not the speaker.

C. They might well sound monotonous.
WRONG: This would not be most likely. There is no passage link between monotony and pitch. Monotony has to do with inflection, not pitch.

D. You could find them difficult to understand.
CORRECT: This would be most likely. The key is "*unusually* low". "[A]n artificially lowered voice can sacrifice intelligibility" (lines 7-8).

23. The author most likely believes that one of the main purposes of speaking, during a face-to-face meeting, should be to:

A. convey a favorable impression.
CORRECT: This is not the most likely belief of the author. Almost the entire passage is about conveying favorable *impressions*, not ideas or concepts. The author admits that "the quality of content accounts for a mere 6%" (lines 24-25) of the overall impression given by the speaker.

B. effectively transmit your ideas.
WRONG: This is not the most likely belief of the author. There is no support for this answer within the passage. The author admits that "the quality of content [i.e. ideas] accounts for a mere 6%" (lines 24-25) of the overall impression given by the speaker.

C. gain leverage.
WRONG: This is not the most likely belief of the author. There is no support for this answer within the passage.

D. communicate as naturally as possible.
WRONG: This is not the most likely belief of the author. For instance, the author would not be a proponent of communicating "as naturally as possible" if you *naturally* were 1. shy, 2. spoke with a high-pitched whine, 3. were 'breathy', etc.

24. The author provides a list of "habits to be avoided" (lines 13-15). Which of the habits would the suggestions in this passage *not* help a speaker to curb?

Notice that, of the four answer choices, Answer C. "pretension" differs in 'kind' from the other choices. In other words, if you had a list of only these four choices with a new question asking, 'Which one of the following items doesn't belong with the others?', you would still choose "pretension".

A. monotony

WRONG: There is information within the passage which would help a speaker to curb "monotony". Speech should be "inflected to suggest the emotions expressed" (line 3).

B. breathiness

WRONG: There is information within the passage which would help a speaker to curb "breathiness". See lines 36-54.

C. pretension

CORRECT: There is no information within the passage which would help a speaker to curb "pretension". Though we may make some assumptions, the passage does not convey how "pretension" is conveyed. Since the symptoms of 'pretension' are not defined, there is little in the way of help which can be gleaned from the passage for the person who suffers from 'pretension'.

D. high-pitched whining

WRONG: There is information within the passage which would help a speaker to curb "high-pitched whining". "To deepen the pitch, speakers should make an extra-deep inhalation before speaking, and then exhale fully as they speak" (lines 10-12).

25. The term *ideal speaking voice* (line 1) refers implicitly to a voice that is:

A. the most pleasant to listen to.

CORRECT: This is not the most strongly implied answer. The author gives us our first clue when describing a voice which might "sacrifice intelligibility, which is *irritating* to listeners. Thus, speakers should experiment to find their *optimal level*, which will be at their lowest intelligible pitch" (lines 7-10). At this point, we might assume that, thus, the "optimal level" or ideal speaking voice is the "least irritating" (Answer C.). However, overall, the implication of the passage is seeking to be the *best*, not (forgive the tortured syntax) the "least" worst of the worst.

B. the most persuasive.

WRONG: This is not the most strongly implied answer. This has to do with 'content'. We can surmise that since "the quality of content accounts for a mere 6%" (lines 24-25) of the overall impression given by the speaker, "most persuasive" is not a characteristic of the ideal speaking voice.

C. the least irritating.

WRONG: This is not implied. The author seeks the *best*, not (forgive the tortured syntax) the "least" worst of the worst.

D. the most natural.

WRONG: This is not implied. For instance, the author would not be a proponent of the idea speaking voice being the "most natural" if your *"most natural"* speaking voice was 1. shy sounding, 2. a high-pitched whine, 3. or 'breathy', etc.

26. Passage information indicates that a person speaking in a high-pitched voice might be doing all of the following EXCEPT:

I. Breathing with their upper lungs

WRONG: This is not an exception. A person speaking in a high-pitched voice might be "breathing with their upper lungs". "Diaphragmatic breathing results in a deeper voice than upper-lung breathing" (lines 40-41).

II. Breathing deeply

CORRECT: This *is* an exception. A person speaking in a high-pitched voice would probably *not* be "breathing deeply". For instance, "To deepen the pitch, speakers should make an extra-deep inhalation before speaking, and then exhale fully as they speak" (lines 10-12).

III. Lying
WRONG: This is not an exception. A person speaking in a high-pitched voice might be "lying". "To suggest credibility, the voice's tone should be pitched as low as is naturally possible... A low pitch is desirable in both genders, since it is popularly associated with truth-telling" (lines 3-7).

A. I only

B. II only
CORRECT: See above answer explanations.

C. III only

D. I and III only

27. Which of the following assertions is most clearly a thesis presented by the author?

A. Speakers can gain by improving his or her speaking voices.
CORRECT: This is clearly a thesis presented by the author. It is presented through the negative aspects associated with having a 'poor' speaking voice, such as "irritating" the listener. It is presented through the examples of "famous news anchors" who had great speaking voices. And, it is presented through statistics within the passage; "everyone who communicates should be aware that their voice is a critical component of their audience's perceptions of them, comprising about 38% of the overall impression imparted by their presentation" (lines 19-22).

B. The tone of the ideal speaking voice should be pitched as low as possible.
WRONG: This is not clearly a thesis presented by the author, and inaccurately paraphrases the passage. Speakers are admonished to find their "lowest *intelligible* pitch" (line 10), and to pitch their voices as "low as it *naturally* possible" (line 5), else they will risk irritating their listeners.

C. What you are saying is more important than how you are saying it.
WRONG: This is not clearly a thesis presented by the author. It is an antithesis. "[T]he quality of content accounts for a mere 6%" (lines 24-25) of the audience's perception of the speaker.

D. Emotional inflections can be an irritating aspect of a speaker's voice.
WRONG: This is not clearly a thesis presented by the author. The ideal speaking voice "*should* be ... inflected to suggest the emotions expressed" (lines 2-3).

28. The ideas discussed in this passage would likely be of most use to:

A. A doctor
WRONG: This is not most likely.

B. A journalist
WRONG: This is not most likely. There is no reason to believe that the ideal speaking voice is of more particular use to a journalist who writes for a living, than to anyone else.

C. A radio show personality
CORRECT: This is most likely. The passage is almost solely on the "ideal speaking *voice*". Unlike the "famous [television] news anchors" used by the author as examples, 'radio show personalities' would not require coaching and advice on their visual appearance.

D. A television commentator
WRONG: This is not most likely. The passage is almost solely on the "ideal speaking *voice*". Despite the author's use of "famous [television] news anchors" which might make this answer seem attractive, clearly, television relies at least somewhat, if not just as much, on visual components and appearance, as it does on the speaking voice. However, beyond mentioning that "appearance accounts for about 50% of the speaker's impact" (lines 23-24), there is no further discussion of these ideas within the passage. Nor is there any advice given regarding improving the "appearance". Thus, this passage would *not* be as com-

pletely useful to a 'television commentator' as it would to a 'radio show personality' to whom his speaking voice is his entire persona.

Passage II (Questions 29–35)

29. The author's discussion of "push" and "pull" factors (line 30) most accurately implies that:

A. "pull" factors compel someone to leave, while "push" factors induce someone to come.
WRONG: This is not the most accurate implication.

B. "pull" factors induce someone to come, while "push" factors also induce someone to come.
WRONG: This is not the most accurate implication.

C. "push" factors require someone to leave, while "pull" factors also compel someone to leave.
WRONG: This is not the most accurate implication.

D. "push" factors compel someone to leave, while "pull" factors induce someone to come.
CORRECT: This is the most accurate implication. "For example, the Fujianese feel compelled ("pushed") to leave because of the area's low standard of living. The poor wages, bad housing, and lack of political freedom can also be seen as "pull" factors, due to the idea that the Fujianese understand that life would be better in other countries" (lines 31-35).

30. Given the information in the passage, if "'artificially' defined" borders (lines 3) were eliminated throughout the world, which of the following outcomes would most likely occur?

A. People would naturally immigrate to areas with higher standards of living.
WRONG: This outcome is not the most likely to occur. Though the author does provide that people do tend to immigrate to areas with higher standards of living, there is no particular reason to believe that elimination of artificially defined borders would increase or decrease this tendency, *without more information* from the question.

B. Nations would encounter less traditional border strife.
CORRECT: This outcome is the most likely to occur. The key here is "traditional" border strife. In contrast to the immigration problems predominantly focused upon in the passage, this type of strife is characterized by nation/states "attempting to secure their borders from invading countries, or even seeking to expand their own territories and acquire additional resources" (lines 60-63). <u>Why?</u> Because they would be separated by natural "geographical" boundaries such as wide rivers, mountain ranges, and oceans. Or, cultural boundaries. One can presume that these are boundaries where people sharing the same cultural characteristics, such as language, religious beliefs, etc., have chosen to live together.

C. Nations would require greater border security measures.
WRONG: This outcome is not the most likely to occur. Natural "geographical" boundaries such as wide rivers, mountain ranges, and ocean would be separating peoples who shared the same cultures.

D. People would live more harmoniously.
WRONG: This outcome is not the most likely to occur. This is possible. However, this answer is vague when compared with Answer B.

31. Which of the following assertions does the author support with an example?

A. Transportation methods used by illegal immigrants are sometimes dangerous.
CORRECT: This assertion is supported with an example. "Within the past year, U.S. officials found three Chinese immigrants in a smuggling boat's sealed cargo container, dead from suffocation" (lines 22-24).

B. Peru and Columbia are seeking to expand their own territories.
WRONG: This assertion is not supported with an example. Presumably because it is not a passage assertion or an assertion of the author's!

C. New Zealand has enacted laws that hasten deportation proceedings.
WRONG: This assertion is not supported with an example.

D. The mission of the Peruvian troops is to keep guerillas and drugs out of Peru.
WRONG: This assertion is not supported with an example.

32. The passage as a whole suggests that in order for a nation to slow the exodus of its inhabitants to other countries, it must:

A. become more attractive to those who are leaving.
CORRECT: This is suggested by the passage as a whole. The information regarding the "push" and "pull" factors suggests that poor conditions in the native country 'push' the immigrant, while higher standards in an adjoining country 'pull' the immigrant. Presumably, by becoming "more attractive to those who are leaving" a nation might slow the leaving of its inhabitants. Though "more attractive" might seem vague, it encompasses a variety of options and changes such as higher pay, better standards of living, better political system, more political freedom, etc., that make this the best answer.

B. abandon the traditional methods of guarding borders.
WRONG: This is not suggested by the passage as a whole. We don't even really know what the "traditional methods of guarding borders" refers to. What was it. How are borders guarded now? "Traditional" refers to what occurred along borders which countries shared, not the methods by which they were guarded.

C. respond in some way to the conflicts arising from border disputes.
WRONG: This is not suggested by the passage as a whole. This really has nothing to do with the inhabitants leaving or staying.

D. answer the challenges set forth by adjoining countries.
WRONG: This is not suggested by the passage as a whole. This really has nothing to do with the inhabitants leaving or staying. One might infer that by "challenges", this answer refers to "economic challenges" and rising to meet the standards of the more attractive neighboring country. However, this is a stretch. "Challenges" might just as well mean challenges to war, or sporting challenges.

33. If the passage information is correct, what inference is justified by the fact that virtually no immigration from West Berlin to adjoining East Berlin occurred, over the 40 years before the period described?

A. Crossing the heavily guarded borders between West and East Berlin was very dangerous.
WRONG: This inference is not justified by the new "fact." There is no passage information to support the premise that more 'heavily guarded borders' would be effective in an effort to prevent or slow immigration. Or, that a "very dangerous" situation would prevent immigration. On the contrary, the example of the suffocated Chinese immigrants indicates that danger would not stop a motivated person from attempting to immigrate.

B. It was understood that life would be better in East Berlin.
WRONG: This inference is not justified by the new "fact." Just as the Fujianese migrated to countries and areas where the conditions were more favorable, if this answer were accurate, the question would provide that 'a great deal of immigration has taken place from West Berlin to East Berlin.'

C. The inhabitants of both 'Berlins' were happy to remain where they were.
WRONG: This inference is not justified by the new "fact." The use of the word 'happy' is simplistic and should alert you that this is probably not the best choice. The passage heavily incorporates economics as causal "push" and "pull" factors. Further, you assume that there is no immigration from East Berlin to West Berlin. Notice that this possibility is left open in the question's factual premise. This is not the best answer.

D. The economic conditions of West Berlin were much more favorable than those of East Berlin.
CORRECT: This inference is justified by the new "fact." Again, it is certainly in the realm of possibility

that there is immigration taking place from East Berlin to West Berlin. But whether this is true or not, this answer incorporates passage economic ideas in advancing a reason why there would be "virtually no immigration from West Berlin to East Berlin."

34. The author implies that which of the following is *not* one of the reasons that Peruvian President Fujimori deployed soldiers to its borders with Columbia?

The question requires an implication of the author's. Note the *absence* of an implication. Notice that this type of question differs dramatically from one reading: "The author implies that Peruvian President Fujimori deployed soldiers to its borders with Columbia for all of the following reasons EXCEPT:"

I. Fujimori is attempting to keep drugs out of his country.

WRONG: The author implies that this is one of the reasons the soldiers were deployed.

II. Fujimori fears that Columbia is seeking to expand its territories.

CORRECT: The author implies that this is not one of the reasons the soldiers were deployed. "Traditionally, these two countries might have been attempting to secure their borders from invading countries, or even seeking to expand their own territories and acquire additional resources. However, Ecuador and Peru are protecting their borders *from rogue drug traffickers and guerillas, not Colombia's government*" (lines 59-64).

III. Fujimori is probably concerned that Columbia wants to acquire additional resources.

WRONG: This is not an implication of the author's. The question requires an implication of the author's, not the absence of an implication. This is not an 'exception' type of question. Notice that this type of question differs dramatically from one reading: "The author implies that Peruvian President Fujimori deployed soldiers to its borders with Columbia for all of the following reasons EXCEPT:"

A. I only

B. II only
CORRECT: See above answer explanations.

C. III only

D. II and III only

35. It seems likely that New Zealand may be suffering less from immigration issues than the United States for which of the following reasons:

I. The U.S. offers higher wages than New Zealand.

WRONG: There is no support for this answer in the passage.

II. New laws, enacted in New Zealand, allow faster deportation proceedings.

CORRECT: This answer choice "seems likely". "Recently, New Zealand attempted to deal with these aliens by enacting new immigration laws, which hasten the process required to deport them" (lines 24-27).

III. Immigrants often do not settle in New Zealand.

CORRECT: This idea not only "seems likely", it is provided in the passage as a reason. "What keeps New Zealand from experiencing an even more profound illegal immigration problem is that the immigrants often do not settle there" (lines 38-40).

A. II only

B. III only

C. II and III only
CORRECT: See above answer explanations.

D. I, II, and III

Passage III (Questions 36-42)

36. What is the author's response to the standard story about the woman who spills hot McDonald's coffee in her lap, sues and gets several million dollars?

Since this story is not actually in the passage, the author is not actually responding to it. However, corollaries can be drawn from within the passage. The author's response can be predicted. This type of question is very similar to others found on the MCAT such as, "The author would argue that …", "On the basis of the passage, it is reasonable to conclude that …", and "The author would be most likely to respond by …".

A. This story does not reflect that compensation is usually insufficient.
CORRECT: This is the most likely response to the "standard story". Remember that the author emphasizes that, "In reality, compensation tends toward *inadequacy*" (line 14). He gives several reasons, one of which is the "hurdle" of proving negligence. Thus, the author would not believe that the woman's case was indicative of most other torts.

B. This story is a good example of just the right amount of compensation.
WRONG: This story is *not* a good example of anything that the author promotes. We certainly have no way of knowing if the compensation was "just the right amount".

C. This story does not reflect that deterrence is costly.
WRONG: This is way outside of the scope of the passage. Deterrence is costly? The only information we have regarding the cost of deterrence is for the defendant. Finally, the question has to do with the woman's *compensation*, not deterrence.

D. In this story, the woman was malicious.
WRONG: This makes little sense given the abbreviated information in the question. It has nothing to do with the passage.

37. Which of the following assertions is the most effective argument *against* the author's opinion that personal injury law cannot satisfactorily compensate and deter "as long as there are administrative costs involved in obtaining an award" (lines 1-7)?

A. These administrative costs are inconsequential.
WRONG: This is not the most effective argument against the author's opinion. First, one must realize that the author has essentially *defined* 'administrative costs' as 'attorney's fees' (lines 6-13). Then, one must ask if it is reasonable to argue that attorney's should be compensated for their work. It is reasonable to argue that anyone should be compensated for their work (we are not saying how much or how little), and thus, yes, it is reasonable that attorney's should be compensated for their work. This answer is not reasonable. Further, if the costs are around "30%" of the awards, that is probably not considered by anyone to be "inconsequential".

B. Attorneys are a necessary part of the judicial system and should be compensated for their work.
CORRECT: This is the most effective argument against the author's opinion. First, one must realize that the author has essentially *defined* 'administrative costs' as 'attorney's fees' (lines 6-13). Then, one must ask if it is reasonable to argue that attorney's should be compensated for their work. It is reasonable to argue that anyone should be compensated for their work (we are not saying how much or how little), and thus, yes, it is reasonable that attorney's should be compensated for their work. Finally, unlike some of the other answer choices, this argument is one that the author has not responded to in the passage. This is the best answer.

C. The administrative costs should be added to the compensation received by the plaintiff.
WRONG: This is not the most effective argument against the author's opinion. The author responds to this idea in the passage. In the example, "If she is paid in full [i.e. administrative costs are added to the full compensation], then the defendant is paying 130% of the actual harm caused, and is *over-deterred*" (lines 11-13). This cannot be the *best* argument because it is answered already in the passage.

D. The administrative costs should be subtracted from the compensation received by the plaintiff.
WRONG: This is not the most effective argument against the author's opinion. The author responds to this idea in the passage. In the example, "If the plaintiff is awarded 100% of the damages suffered, she only receives compensation for 70% of her injuries" (lines 9-11). This cannot be the *best* argument because it is answered already in the passage.

38. The passage indicates that its author would NOT agree with which of the following statements?

A. Tanker companies are a good example of defendants who are under-deterred.
WRONG: There is no basis from which to judge whether the author would agree or not agree with this answer. The story of the tanker company is a "permutation on actual events"; it is a fabrication.

B. Negligence on the part of the defendant is generally not difficult for the plaintiff to prove.
CORRECT: The author would not agree with this statement. The author argues in several ways the "hurdle" (line 19) of proving negligence. Besides that fact that 'hurdle' connotes difficulty in surmounting, the author also tells us that "compensation tends toward inadequacy" (line 14) specifically because negligence is so hard to prove.

C. The costs associated with suing and defending against suits can be tremendous.
WRONG: There is no basis from which to judge whether the author would agree or not agree with this answer. The only dollar amounts provided are in the last paragraph. The idea of "tremendous" costs is relative and opinions between Bill Gates and a homeless person might vary.

D. In many situations, over-deterrence results in primarily economic ramifications.
WRONG: The author would probably agree. Over-deterrence "results in a waste of resources" (lines 50-51).

39. Assume that since the 9-11 terrorist attacks on the World Trade Center (WTC) buildings, all lawsuits have been settled by the WTC insurance companies, who have now mandated that they will no longer insure any building in the world that is over five stories tall. The author's comments suggest that this situation could reasonably be interpreted as evidence that:

A. the insurance companies were over-deterred.
CORRECT: A reasonable interpretation of the situation, based upon the author's information in the passage could lead one to this conclusion. This idea parallels the 'tanker trial' example offered by the author. Here, after settling the lawsuits from the WTC torts, the insurance companies have mandated a very extreme policy by anyone's estimation. Perhaps the WTC people were at fault which resulted in huge payments to the plaintiffs. Nevertheless, it seems that the WTC insurance companies were, in fact, "over-deterred", resulting in "a waste of resources" (lines 50-51).

B. the insurance companies were under-deterred.
WRONG: This is not a reasonable interpretation of the situation, based upon the author's information in the passage. "Under-deterrence" results in a defendants "effectively hav[ing] no incentinve" to change or prevent future harms. Here, after settling the lawsuits from the WTC torts, the insurance companies have mandated a very extreme policy by anyone's estimation.

C. the plaintiffs were under-compensated.
WRONG: This is not a reasonable interpretation of the situation, based upon the author's information in the passage. This is possible, but very unlikely. There *might* have been so many plaintiffs that their rather *small* compensatory damage awards simply overwhelmed the insurance companies. However, it is not probable. It is not even as probable as Answer D, which is also not the best answer. Review the explanations for Answer D and Answer A.

D. the plaintiffs were overcompensated.
WRONG: This is not a reasonable interpretation of the situation, based upon the author's information in the passage. This is possible. With a poor understanding of the author's concepts from the passage, this answer might seem to go hand-in-hand with Answer A; it might seem just as "correct". It is not. We have no way of knowing if the plaintiffs were overcompensated. Every one of the plaintiffs *might* have been

making billions of dollars a year, yet rendered helpless quadriplegics requiring full-time 24-hour care which was *not* completely compensated by the lawsuits. There *might* have been so many plaintiffs that their rather small compensatory damage awards simply overwhelmed the insurance companies. We have no way of knowing based upon the information provided. Consider that, as mentioned, it is conceivable that over-deterrence and under-compensation occur simultaneously, to the benefit of no one. This is not the best answer.

40. Suppose that a study found that police agencies routinely set aside large amounts of money in their yearly budgets, which they expect to pay out in lawsuits against their agency. Which of the following statements is an assumption of the author about the effects of lawsuit awards that would be called into *question*?

The correct answer *must* satisfy *two* criteria. It must 1. be an assumption of the author about the effects of lawsuit awards, and 2. be called into question by the given supposition.

A. Simply proving negligence can be a very costly process.
WRONG: This statement is not a clear assumption of the author, and is not clearly called into question by the supposition.

B. Many people will not sue because the process is too costly.
WRONG: This is an implication of the author's since the idea was brought up in an example (lines 72-73), though the word "many" makes it highly questionable. Further, the supposition clearly states that the 'set aside' money is *not* for *defending* against lawsuits, but in order to "pay out" the awards.

C. If a plaintiff receives full compensation and administrative costs, the defendant is over-deterred.
CORRECT: This statement is very clearly an assumption of the author's (lines 11-13). Further, this statement *is* called into question by the supposition. If the police agency "routinely" sets aside/budgets the money it will have to pay out in awards, then it seems that this is just a business-as-usual approach. The author's definition of over-deterrence results in a "change". There is an "incentive to prevent [harm]".

D. Depending upon the size of the award, a defendant police agency might not be deterred at all.
WRONG: This is not clearly an assumption of the author, since it specifies police agencies, though it does somewhat parallel the idea that in situations where there is no award "defendants effectively have no incentive to prevent small harms" (lines 75-76). However, this statement is actually *supported* by the supposition. It is certainly not questioned by it.

41. Which of the following conclusions can justifiably be drawn from the experience of the tanker company's counsel mentioned in the passage?

A. Good economists make for poor attorneys.
WRONG: This conclusion cannot be justifiably drawn from the experience of the counsel (i.e. attorney). This answer is attractive because of the author's reference to "the company's counsel — a good economist, but a poor lawyer" (lines 29-30). However, there is no indication that *being* a good economist made this person a poor attorney.

B. Costs should never be considered prohibitive where safety is concerned.
WRONG: This conclusion cannot be justifiably drawn from the experience of the counsel (i.e. attorney). This is outside the scope of the question. The question does not ask about the tanker *incident in general*, but the tanker company's counsel *specifically*.

C. Toxic materials should not be shipped through residential neighborhoods.
WRONG: This conclusion cannot be justifiably drawn from the experience of the counsel (i.e. attorney). This is outside the scope of the question. The question does not ask about the tanker *incident in general*, but the tanker company's counsel *specifically*.

D. Honesty is not always the best policy for an attorney.
CORRECT: This conclusion can be justifiably drawn from the experience of the counsel (i.e. attorney). The author has already informed the reader that the tanker lawyer is a "poor lawyer". Thus, his actions would

not be an example for others to follow. The lawyer "*admits* safer tankers were available, but the cost is prohibitive. After extensive cost-benefit analyses, he says, the company found it cheaper just to pay victims for their losses, as it now offers to do" (lines 30-34). The author emphatically (!) announces that the company would be lucky to "escape punitive damages" (line 35). Apparently, the attorney was simply being honest, which (at least from the standpoint of his clients) was probably not the best idea.

42. The author argues that, "Potentially, the most promising development in tort (personal injury) law has been the advent of strict liability" (lines 54-56). These beliefs imply that:

Note that you must understand what "strict liability" is.

A. the use of strict liability has become increasingly popular for defendants.
WRONG: This is not implied by the beliefs in the author's argument.

B. the uses of strict liability should remain limited in scope.
WRONG: This is not implied by the beliefs in the author's argument. According to the passage, strict liability *is* limited in scope. The author believes that, "Unfortunately, the *application* of strict liability is severely constrained by legal doctrine, which limits its application to a small range of 'unusually hazardous activities'" (lines 59-63).

C. the author approves of waiving the requirement for proof, where carelessness is evident.
CORRECT: This answer is implied by the beliefs in the author's argument. "Potentially, the most promising development in tort (personal injury) law has been the advent of strict liability, which waives plaintiffs' need to prove the defendant's carelessness in certain instances where the carelessness is obvious, or could have resulted from no factor other than negligence" (lines 54-59).

D. the author approves of compensation where carelessness is evident.
WRONG: This is not implied by the beliefs in the author's argument. This answer may be accurate information according to the passage. However, that differs from specifically answering the question. You cannot simply choose an answer that provides accurate information to the question. It must be accurate *and* the most responsive answer to the question.

ANSWERS & EXPLANATIONS FOR 30-MINUTE IN-CLASS EXAM 3

Passage I (Questions 43–49)

43. The passage suggests that its author would probably *disagree* with which of the following statements?

A. It is possible that Chris participated in "male" activities in order to be considered male.
WRONG: The author would *not* disagree with this statement. From paragraph (lines 45-59) the author discusses the "two sides of the issue"; this statement being one of the sides. "Almost certainly, there is a complex interaction between the two" (lines 56-57).

B. It is possible that Chris naturally participated in "male" activities.
WRONG: The author would *not* disagree with this statement. From paragraph (lines 45-59) the author discusses the "two sides of the issue"; this statement being one of the sides. "Almost certainly, there is a complex interaction between the two" (lines 56-57).

C. Chris was not confused about her identity.
WRONG: The author would *not* disagree with this statement. "Chris was in no way confused about her identity" (lines 21-22).

D. Most cultures have clearly defined gender roles.

CORRECT: The author would disagree with this statement. The word, "most", in this statement renders it disagreeable. The author alludes to *other* cultures where Chris might not have felt the need "to identify herself as distinctly male" (lines 40-41). The culture/society wherein Chris lives is clearly defined as "our" (line 5 and 77) culture/society, implying that there are others. There is simply no way of quantifying this statement. This answer choice can also be arrived upon through process of elimination since the other answers the author would clearly *not* disagree with.

44. Implicit in the passage is the assumption that:

I. one should be happy in one's "natural" state.

WRONG: This is not implied. Though we may strive for "an acceptance of ourselves in a 'natural' state" (line 64), and psychotherapy may help us to arrive at this point, there is not value-judgment-type "should" or "should not" implication in the passage from which to choose this answer. The author seems to have no problem with the idea that Chris would want a sex change operation.

II. one can be well-adjusted, yet unhappy with one's "natural" state.

CORRECT: This is clearly implied. We know from the passage that Chris "is described as a relatively well-adjusted individual" (line 24). Additionally, we know that Chris felt "she was a man in a woman's body" (line 17). "Unhappiness" can be used to describe this state because of the information at lines 67-69 "The effect of a physical sex reassignment operation on Chris' happiness cannot be foretold with complete certainty". Apparently, Chris "happiness" was an issue.

III. one's perception of self is most important.

WRONG: This is not implied. This assumption is defined by the quote from the passage that "through psychotherapy, one learns that one may not necessarily have to change oneself as much as *one's perception of self*" (lines 65-67). There is no indication or implication in the passage that this is "most important", or *more* important than *changing* the "natural state" to fit our perception of self.

A. I only

B. II only

CORRECT: See above answer explanations.

C. III only

D. I and III only

45. The author of the passage would be most likely to agree with which of the following ideas expressed by other psychologists?

A. A DSM IV 'disorder' may not actually be a disorder at all.

WRONG: The author would not *most likely* agree with this idea. This idea is perhaps a second best choice. But it is much more vague than Answer C. It is clear from the passage that the DSM is a *reflection* of societal norms. In *this* society, in "*our*" society, even the author admits that Gender Identity Disorder *is* a disorder.

B. The DSM IV is a poor descriptor of abnormal behavior and desires since it is easily influenced by societal norms.

WRONG: The author would not *most likely* agree with this idea. Though it seems that the DSM is influenced by societal norms, there is no indication that it is "easily" influenced, or that overall it is a "poor descriptor" of abnormal behavior. It may actually be a very accurate descriptor of behavior that society has determined is abnormal.

C. Some DSM IV 'disorders' are simply an attempt to characterize socially abnormal behavior and desires.

CORRECT: The author would most likely agree with this idea. The author argues that in another culture, Chris' disorder *might* not even be a disorder. "The fact that Gender Identity Disorder exists in the DSM IV [the official handbook of psychiatric disorders] as a diagnosis is an admission on the part of psychologists that our society has clearly defined gender roles. These contribute to what it is generally considered "normal"" (lines 2-7). In *this* society, in "*our*" society, even the author admits that Chris' is considered to have a disorder.

D. Behavior and desires must fall within the parameters of the DSM IV to be considered normal by society.
WRONG: The author would not most likely agree with this idea. First, behavior falling "*within*" the parameters of the DSM IV is considered *abnormal*, not normal. Second, it is societal norms which the author believes have determined the DSM parameters. Not the other way around.

46. The author hints that the fact that Christ is well-adjusted indicates that her "uneasiness with her assigned sex" (lines 30-31):

A. is a problem which should be overcome through psychiatry.
WRONG: This is not hinted at by the author. This is too strong. The author admits that the treatment of Chris is still a "*question*" (line 58). There is no implication that she "should" overcome her problem through psychiatry. "Psychologically healthy individuals" accept themselves in a "natural" state (lines 63-64). However, the author seems to believe that the "natural" state may just as easily be altered, as one's perception of self.

B. is due to the culture she lives in.
CORRECT: This is hinted at by the author. Paragraph 33-43, in its entirety, *strongly* argues this answer; beginning with, "in this case, many psychologists may believe society is the culprit" (lines 34-35).

C. can be overcome through surgery.
WRONG: This is not hinted at by the author. This is too strong. The author admits that the treatment of Chris is still a "*question*" (line 58). There is no implication that she "can" overcome her problem through surgery.

D. is a basic personality defect.
WRONG: This is not hinted at by the author. There is no support for this answer in the passage.

47. Suppose it is discovered that prescription medication allows Chris to become somewhat more comfortable with her "natural" state. Does this discovery support the author's argument?

What is the author's argument? This passage is *not* a completely objective representation of a psychological case study. His main argument is that "society is the culprit". "One *must* ... ponder the age-old question of society as the cause" (lines 31-32).

A. Yes; it confirms it.
WRONG: This discovery does not support the author's argument. For this to be a correct answer choice, the author would have to be arguing that there is a psychological or neurological "causal factor". However, the author finds that Chris is "well-adjusted" and dismisses this idea and posits that, "One *must* ... ponder the age-old question of society as the cause" (lines 31-32).

B. No; it does not affect it.
CORRECT: This discovery supports the author's argument. This passage is *not* a completely objective representation of a psychological case study. His argument is that "society is the culprit". "One *must* ... ponder the age-old question of society as the cause" (lines 31-32). He implies in paragraph 33-43 that in another culture, Chris might not even be suffering from a disorder. Her unhappiness can be linked to society's unwillingness to accept her. Therefore, even if drugs help her become "*somewhat*" more comfortable with her "natural" state, the author's argument is not affected by this discovery.

C. No; it weakens it.
WRONG: This discovery does not support the author's argument. The author's argument that "society is the culprit", is *not* weakened by an unspecified *dosage* and *type* of "prescription medicine" rendering Chris only "*somewhat* more comfortable". The supposition in the question is vague enough to assume that she was on a very strong dosage of a very powerful psychotropic drug. This is not the best answer.

D. No; it disproves it.
WRONG: This discovery does not support the author's argument. This answer and Answer A are the poorest of the four choices. See the explanations to Answer B.

48. The author admits that, "The effect of a physical sex reassignment operation on Chris' happiness cannot be foretold with complete certainty" (lines 67-69). The author most likely believes that:

I. it is just as likely that psychotherapy would help Chris to change her perception of self.
WRONG: This is not what the author most likely believes. The author believes that Chris is "well-adjusted" and that "society is the culprit" (line 35). The author's nod to psychotherapy is lukewarm at best and provided only in the spirit of seeming to be objective (lines 64-67).

II. in our society, in the body of a woman, Chris will not be happy.
CORRECT: The author most likely believes this. This passage is *not* a completely objective representation of a psychological case study. His argument is that "society is the culprit". "One *must* ... ponder the age-old question of society as the cause" (lines 31-32). He implies in paragraph 33-43 that in another culture, Chris might not even be suffering from a disorder. Her unhappiness can be linked to society's unwillingness to accept her acting as a man in the body of a woman.

III. *a "sex reassignment operation" would make Chris happier.
CORRECT: The author most likely believes this. This answer is tantamount to Answer II. See the above explanation.

A. I only

B. II only

C. III only

D. II and III only
CORRECT: See the above answer explanations.

49. The author's attitude toward "our" societal norms is most accurately described as:

A. favorable.
WRONG: This is not an accurate description.

B. neutral.
WRONG: This is not an accurate description.

C. distrustful.
WRONG: This is not an accurate description because it is not strong enough. Further, "trust" or "distrust" are poor descriptors for the author's attitude in the passage. See the explanation for Answer D.

D. disapproving.
CORRECT: This passage is *not* a completely objective representation of a psychological case study. The author's argument is that "society is the *culprit*" (line 35). "One *must* ... ponder the age-old question of society as the cause" (lines 31-32). He implies in paragraph 33-34 that in another culture, Chris might not even be suffering from a disorder. Her unhappiness can be linked to society's unwillingness to accept her acting as a man in the body of a woman. "Our" societal norms are described disparagingly as a "neatly constructed gender dichotomy" (lines 77-78).

Passage II (Questions 50–56)

50. The author of the passage would be most likely to agree with which of the following ideas expressed by other technology theorists?

A. *Soliciting further ideas and diverse ideas is not always wrong.
CORRECT: The author would be most likely to agree with this idea. This answer is restating the rather stilted statement from the passage, "No one can argue that giving additional and different perspectives is inherently bad" (lines 54-56).

B. Obtaining additional and differing perspectives is always beneficial.
WRONG: The author would not be likely to agree with this idea. The word "always" in this idea is extreme. This idea is certainly not the same as the rather stilted statement from the passage, "No one can argue that giving additional and different perspectives is inherently bad" (lines 54-56). At most the author offers that "it *can* also be argued that more information is good" (line 52). However, conspicuous by its absence is the necessary "*always*" good.

C. For information to be of value, it must challenge our human values.
WRONG: The author would not be likely to agree with this idea. This is an antithesis. According to Nardi and O'Day, whom the author seems to agree with, "If technology doesn't assist in promoting … [our human] values, then it cannot be considered useful" (lines 18-20).

D. A great deal of the information which we receive is useless because it does not pertain to us.
WRONG: The author would not be likely to agree with this idea. He certainly would not be as likely to agree with this idea as the idea given in Answer A. Answer D is actually an idea of Postman's, not the author's; "[Postman] goes on to say that much of the information we receive does not pertain to us" (lines 44-45). First, the author does not agree with some of what Postman posits (lines 50-56). Further, even Postman does not say that this renders the information "useless". He refers to it as "at best an irrelevant distraction" (lines 45-46).

51. The passage argument suggests that information recipients might benefit from:

I. a more rapid flow of information.
WRONG: This is not suggested. Though the author does not completely agree with everything that Postman says, even the author seems to think that recipients are receiving all the information that they can handle at this point.

II. thoroughly examining their information.
CORRECT: This is suggested. The author argues that "information is useless unless it is thoroughly examined" (line 57) and "we as a society must consciously attempt to not only obtain information, but to analyze it for meaning" (lines 60-63). There are no counterpoints offered in the passage to this argument.

III. limiting non-pertinent information.
WRONG: This is not suggested. The idea is only implied as counterpoint to the main passage arguments. Only Postman, whom the author does not completely agree with, says that "much of the information we receive does not pertain to us [and that] this may create a problem …" (lines 44-47). Yet, the reader is left only to assume that Postman would go on to 'limit non-pertinent information'; this is not stated. However, the author then argues (lines 50-56) that Postman's view is too extreme. The author sees Postman as wanting to "reduce the amount of information available" (lines 58-59) and is against this.

A. I only

B. II only
CORRECT: See above answer explanations.

C. III only

D. II and III only

52. In describing Neil Postman's "information glut", the author uses the example of "images of violence in the Middle East [that] do not affect us" (line 48). The author's point is that:

A. the Middle East is at best an irrelevant distraction.
WRONG: This is not the author's point. Perhaps Postman would suggest this. However, the author, who does *not* want to "reduce the amount of information available" (lines 58-59), would not agree.

B. the Middle East is poorly understood by us.
WRONG: This is not the author's point, and neither is it suggested in the passage.

C. these images should affect us.
CORRECT: This is the author's point. The author prefaces his example of the Middle East categorizing it as a "*problem*" (line 47) that we are *not* affected by these images.

D. these images should not be so easily accessible.
WRONG: This is not the author's point. The author would not limit "access" to information, which is tantamount to limiting the amount of information available. Instead, he would better tailor the information "to individual recipient's needs" (line 60).

53. If the following statements are true, which would most *weaken* the argument of the author?

A. The degree to which technology determines human values is questionable.
WRONG: This statement would not most weaken the argument of the author. This answer/statement admits that technology determines human values. It is only a question of the "degree" of change. To paraphrase the author: Human values "are not static", but change according to certain factors, one of the most significant being technology. Based solely upon passage information and this statement it is difficult to determine if there is any disagreement at all.

B. Information must be analyzed for its relevance to human values.
WRONG: This statement would not most weaken the argument of the author. This is the thesis of the passage.

C. Human values are unchanging.
CORRECT: This statement would most weaken the argument of the author. The author argues that "Human values ... are not static [i.e. unchanging]" (lines 1-2). His thesis rests upon this idea since he considers technology to be a "significant" factor affecting human values.

D. Human values are culturally dependent.
WRONG: This statement would not most weaken the argument of the author. It would matter little to the author's arguments whether they were culturally dependent or not.

54. The author argues, "more information is good" (line 52). Unlike Neil Postman, the author does not consider which of the following to be a factor that might limit the usefulness of information?

The correct answer must be one that Postman considers a "limiting factor", but the author does not.

A. Space
WRONG: This is not implied as a limiting factor by Postman, and is not alluded to by the author.

B. Distance
WRONG: This is not implied as a limiting factor by Postman, but is offered as an example of Postman's 'information glut' idea *by the author*. Providing this example *of Postman's ideas*, the author writes "for example, images of violence in the Middle East do not affect us because they are ... too far away to influence us" (lines 47-50). Further, the author never responds to this idea of distance as a limiting factor.

C. Frequency
WRONG: This is not implied as a limiting factor by Postman, and is not alluded to by the author. "Frequency" cannot be assumed to be the same as 'quantity', or 'amount'.

D. Time

CORRECT: This is implied as a limiting factor *by Postman*, and is alluded to by the author, who does not seem to agree that it might limit the usefulness of information. Postman "states that the rapid flow of information allowed by automation and Internet/email provides vast amounts of data to many users *quickly* and *simultaneously*, but doesn't allow humans to understand its meaning *fast enough to keep up*" (lines 40-44). In other words, they don't have time to process the information. However, the author apparently doesn't think that this is a problem. He seems to believe that people generally have all the time in the world. "By giving people *more* information [*which admittedly takes more time for them to process it*], they have a wider basis from which to make decisions and form opinions. No one can argue that giving additional and different perspectives is inherently bad [*except that we usually don't have the time*]" (lines 52-56).

55. According to the passage, one drawback of technology is that it can:

A. supersede all other values.

WRONG: This is not a drawback of technology. First, technology is not implied to be a "value". The "other" in this answer clearly implies that technology is a value. Though the passage provides that, "Technology may force people away from their own values [which] creates a disturbing image of a world where technology itself becomes the center of a society *without values*" (lines 35-39). This idea still does not make technology a value.

B. be used to destroy human values.

WRONG: This is not a drawback of technology. Though the passage provides that, "Technology may force people away from their own values [which] creates a disturbing image of a world where technology itself becomes the center of a society *without values*" (lines 35-39), there is no inference that "the technology can be used to destroy human values". The two do not have the same meaning.

C. subtly change people's values.

CORRECT: This is a drawback of technology. Notice the use of the 'softener' "subtly" which makes this answer even more palatable. We know that technology is a "significant factor" that *can* change people's values (lines 1-5). "Technology may force people away from their own values" (lines 35-38). This is a drawback illustrated by Ellul's "disturbing image" (lines 37-39). The passage provides that technology should fit with the values and not vice versa. One way to ensure this is the "three-pronged format" posited by Nardi and O'Day.

D. become an irrelevant distraction.

WRONG: This is not a drawback of technology. It is "*information*" that can become an "irrelevant distraction rather than the useful news it was intended as" (lines 46-47).

56. Suppose it could be established that technology is most efficient when it performs its function unobtrusively; without being noticed. The author of the passage would be most likely to respond to this information by:

This is a classic MCAT question. Consider your best answer choices those that would allow the author to either *support* or, in the worst case, *resurrect* or *rehabilitate* his main arguments and thesis. No author would be likely to admit he was completely wrong or abandon his thesis.

A. suggesting that this determination ratifies his thesis.

WRONG: The author would not be most likely to respond in this way. The supposition in the question is *somewhat* in opposition to his thesis. The author agrees with Nardi and O'Day that, "If the technology is taken for granted, the actual purpose for using it can become clouded" (lines 24-25). The author would be required to do some explaining to resurrect his ideas.

B. proposing that we must still analyze the information for meaning.

WRONG: The author would not be most likely to respond in this way. Though the entire second half of the passage is about information, this is tangential to the overall thesis on technology. It begins with the example of 'information glut' and snowballs. However, this answer is not really responsive to the supposition. *Information* technology would be only one small aspect of *technology*.

C. asserting that efficiency is usually degraded when a technology is taken for granted.
WRONG: The author would not be most likely to respond in this way. The supposition in the question is *somewhat* in opposition to his thesis. The author agrees with Nardi and O'Day that, "If the technology is taken for granted, the actual purpose for using it can become clouded" (lines 24-25). However, it is a leap of logic to assume that what the author and these analysts specifically mean is that "efficiency is usually degraded".

D. explaining that we must still remain aware of the technology and its intended purpose.
CORRECT: The author would be most likely to respond in this way. The supposition in the question is *somewhat* in opposition to his thesis. The author agrees with Nardi and O'Day that, "If the technology is taken for granted, the actual purpose for using it can become clouded" (lines 24-25). Therefore, by responding in the fashion of Answer D, the author has responded to the seemingly opposing-supposition without losing ground in his own arguments.

Passage III (Questions 57–63)

57. Assume that several others who had attended the same opera were interviewed. If they remarked that Margaret sang with tremendous passion, these remarks would *weaken* the passage assertion that:

A. Mephisto had rendered her irresistible to Faust.
WRONG: This answer is not a "passage assertion", which is a requirement for a correct answer.

B. the lyrics which she sang were "throwaway".
WRONG: This answer is a "passage assertion". However, it is not weakened by the assumption in the question. The remarks have to do with the "tremendous passion" of Margaret's singing, not the "*lyrics*". This answer is not as responsive to the question as Answer C.

C. the young woman of Faust's dreams sang without emotion.
CORRECT: This answer is a "passage assertion". First, the young woman of Faust's dreams *is* a "commoner named Margaret" (lines 51-52). Second, the passage asserts and alludes to "Margaret's *emotionless* singing" (line 57).

D. the young woman of Faust's dreams was a commoner.
WRONG: This answer is a "passage assertion". The young woman of Faust's dreams *is* a "commoner named Margaret" (lines 51-52). However, the remarks do not weaken this assertion because they have no relation to it.

58. On the basis of the passage, it is reasonable to conclude that:

A. Faust lost his soul to the devil.
WRONG: This is not a reasonable conclusion. There is no sure way of knowing this from the passage.

B. the "country fair thronged with revelers" was not in the book.
WRONG: This is not a reasonable conclusion. There is no sure way of knowing this from the passage.

C. the operatic interpretation differed from the book.
CORRECT: This is a reasonable conclusion. There are several instances which allude to and support this conclusion: "In the director's vision …" (lines 11-12), and "A chorus presents the essential plot of Faust, *reduced* from its *several incarnations* …" (lines 21-22), and "… this version presents …" (line 25 and line 48), and "a theme which is not really congruent with the Faust myth …" (lines 55-56).

D. the author did not enjoy the performance.
WRONG: This is not a reasonable conclusion. This is certainly arguable. Though the author indicates that at Scene II the opera was not particularly to his liking, he begins the passage by asserting that "the New York City Opera production of "Mephistopheles" deserves high marks for visual excellence" (lines 1-3).

59. According to the passage, the author felt that the New York City Opera production of "Mephistopheles":

A. was plagued with a poor characterization of Mephisto.
WRONG: This is not supported by passage information. The author actually seems to have liked Mephisto's characterization.

B. suffered from noticeable weaknesses beginning in Scene II.
CORRECT: "In Scene II, … the play devolves into … very forgettable arias [and] … a theme which is not really congruent with the Faust myth" (lines 50-56).

C. was enhanced by Dr. Faust's singing.
WRONG: This is not supported by passage information. There are no allusions to Faust's singing.

D. could have been improved in Scene III.
WRONG: This is not supported by passage information. There are no allusions to Scene III.

60. According to the passage, the author seems to have most enjoyed:

A. the music of the opera.
WRONG: This is not supported by passage information. It is not clear if the author really enjoyed the music. He said that the music was "only slightly corny" (line 6). He certainly did not like the music as much as the "visual excellence" of the production.

B. the singing of the opera .
WRONG: This is not supported by passage information. The author cared little for Margaret's "emotionless singing", for instance.

C. the plot of the opera.
WRONG: This is not supported by passage information. All this we know from the passage, is that this operatic version differed from the book and other versions which the author was familiar with. Further, at Scene II, and at the "Walpurgis night" scene, the author seems not have cared for the way in which the opera was presented.

D. the images of the opera.
CORRECT: "… the New York City Opera production of "Mephistopheles" deserves high marks for visual excellence" (lines 1-3).

61. Which of the following does the author suggest was a component of the original "Faust myth" (lines 55-56)?

I. *Reason triumphing over feeling

CORRECT: This is clearly suggested. From the passage we know that in the opera "Faust and Margaret sing very forgettable arias about the supremacy of feeling over reason, a theme which is not really congruent with the Faust myth" (lines 53-56).

II. A more evil Mephisto

WRONG: This is not clearly suggested. The second paragraph, for one, describes the character of Mephisto in the operatic production, and does give the impression that he has been portrayed/characterized differently in other versions. However, the author gives no indication that "a more evil Mephisto" "was a component of the original 'Faust myth'". There is not enough specificity in the passage information for one to draw this conclusion.

III. A more powerful God

WRONG: This is not clearly suggested. This answer *seems* attractive because of the passage statement, "Thus, this version presents temptation as essentially a wager, or struggle, between God and the Devil (which, at one time, was a remarkably blasphemous notion, as it contradicted the dogma that God is all-powerful over evil)" (lines 25-29). However, it is not at all clear that the phrase "at one time" refers to the "Faust myth" or about people's attitudes and belief in God *in general*. This is not the best answer.

A. I only
CORRECT: See the above answer explanations.

B. II only

C. III only

D. II and III only

62. According to the passage, through what primary means is the fundamental plot transmitted to the audience?

A. Via visual imagery
WRONG: This is not described as a means of transmitting the 'fundamental' plot to the audience.

B. Via Faust's musings
WRONG: This is not described as a means of transmitting the 'fundamental' plot to the audience.

C. Via the chorus
CORRECT: This is *specifically* described as a means of transmitting the 'fundamental' plot to the audience. "*A chorus presents the essential plot* of Faust, reduced from its several incarnations" (lines 21-22).

D. Via Mephisto
WRONG: This is not described as a means of transmitting the 'fundamental' plot to the audience.

63. Regarding the devil's bargain with Faust, the passage strongly implies that:

A. it is Faust who got the better deal.
WRONG: This is not implied.

B. it is the devil who got the better deal.
WRONG: This is not implied.

C. in other versions, the bargain was with Margaret.
WRONG: This is not implied.

D. in other versions, it is Faust who does the bargaining.
CORRECT: This is implied. "In this version of Faust, it is Faust who seizes the devil's bargain …" (lines 47-48).